The Topology of CW Complexes

HALMOS, PAUL R.—Measure Theory
JACOBSON, NATHAN—Lectures in Abstract Algebra
Vol. I—Basic Concepts *Vol. II*—Linear Algebra
Vol. III—Theory of Fields and Galois Theory
KLEENE, S. C.—Introduction to Metamathematics
LOOMIS, LYNN H.—An Introduction to Abstract Harmonic Analysis
LOÈVE, MICHEL—Probability Theory, 3rd Edition
KELLEY, JOHN L.—General Topology
ZARISKI, OSCAR, and SAMUEL, PIERRE—Commutative Algebra, Vols. I and II
GILLMAN, LEONARD, and JERISON, MEYER—Rings of Continuous Functions
RICKART, CHARLES E.—General Theory of Banach Algebras
KELLEY, J. L., NAMIOKA, ISAAC, and CO-AUTHORS—Linear Topological Spaces
SPITZER, FRANK—Principles of Random Walk
SCHENKMAN, EUGENE—Group Theory
NACHBIN, LEOPOLDO—The Haar Integral
KEMENY, JOHN G., SNELL, J. LAURIE, and KNAPP, ANTHONY W.—Denumerable Markov Chains
HÖRMANDER, LARS—An Introduction to Complex Analysis in Several Variables
SARIO, LEO, and NOSHIRO, KIYOSHI—Value Distribution Theory
HALMOS, PAUL R.—A Hilbert Space Problem Book
RODIN, BURTON, and SARIO, LEO—Principal Functions
VARADARAJAN, V. S.—Geometry of Quantum Theory, Vol. I
GRÄTZER, GEORGE—Universal Algebra
LUNDELL, ALBERT T. and WEINGRAM, STEPHEN—The Topology of CW Complexes

A series of advanced text and reference books in pure and applied mathematics. Additional titles will be listed and announced as published.

The Topology of CW Complexes

Albert T. Lundell
University of Colorado

and

Stephen Weingram
Purdue University

Van Nostrand Reinhold Company

NEW YORK CINCINNATI TORONTO LONDON MELBOURNE

To our wives

ISBN 978-1-4684-6256-2 ISBN 978-1-4684-6254-8 (eBook)
DOI 10.1007/978-1-4684-6254-8

Van Nostrand Reinhold Company Regional Offices:
Cincinnati, New York, Chicago, Millbrae, Dallas

Van Nostrand Reinhold Company Foreign Offices:
London, Toronto, Melbourne

Softcover reprint of the hardcover 1st edition 1969

Library of Congress Catalog Card Number 68 26689

Published by Van Nostrand Reinhold Company
450 West 33rd Street, New York, N. Y. 10001

Published simultaneously in Canada by
D. Van Nostrand Company (Canada), Ltd.

10 9 8 7 6 5 4 3 2 1

Preface

Most texts on algebraic topology emphasize homological algebra, with topological considerations limited to a few propositions about the geometry of simplicial complexes. There is much to be gained however, by using the more sophisticated concept of cell (CW) complex. Even for simple computations, this concept ordinarily allows us to bypass much tedious algebra and often gives geometric insight into the homology and homotopy theory of a space. For example, the easiest way to calculate and interpret the homology of $\mathbf{CP}^n$, complex projective n-space, is by means of a cellular decomposition with only $n+1$ cells. Also, by a suitable construction we can "realize" the singular complex of a space as a CW complex and perhaps thus give a more geometric basis for some arguments involving singular homology theory for general spaces and a more concrete basis for singular homotopy type. As a final example, if we start with the category of simplicial complexes and maps, common topological constructions such as the formation of product spaces, identification spaces, and adjunction spaces lead us often into the category of CW complexes. These topics, among others, are usually not treated thoroughly in a standard text, and the interested student must find them scattered through the literature.

This book is a study of CW complexes. It is intended to supplement and be used concurrently with a standard text on algebraic topology. Chapters I and II present the elementary theory of CW complexes. In Chapter III we discuss regular and semisimplicial CW complexes, special types which can be subdivided into simplicial complexes. Chapter IV concerns the homotopy type of a CW complex. In Chapter V, we develop a homology theory for CW complexes which is suitable for calculations in many cases. The exposition is intended to be detailed in Chapters I and II, so that a student can read and absorb the material by himself with minimal effort. The later chapters omit some of the easier details. These are to be regarded as exercises.

In Chapters I through IV, in order not to overlap standard texts and to make the material "elementary," we have avoided algebraic methods and statements, unless the situation becomes awkward or unless such procedure does the reader a disservice because results are usually stated in algebraic terms. In particular, algebraic techniques have not been used in the first two chapters. Passages in which a knowledge of general topology or the very simplest elements of homotopy theory are not sufficient are preceded by an asterisk. Although we consider briefly the traditional material on simplicial complexes, usually by a simple statement to fix notation, we assume that the reader has some familiarity with it. In the same spirit, in Appendices I and II we discuss often-used material on paracompactness and neighborhood extension and retraction properties.

In Chapter V we assume a knowledge of the axioms of singular homology theory and their development from the singular chains as well as an acquaintance with the elementary propositions and terminology of homological algebra. Since our objective in this chapter is to develop a theory that is more amenable to calculation than the singular theory, we have given references to the proofs of the deeper theorems about singular theory. Several applications of the theory are given, among which are the calculation of the homology of several well-known CW complexes.

Throughout the book cross references are given as follows: II.3.5 will refer to Chapter II, section 3, formula or theorem 5; 3.5 will refer to section 3, formula or theorem 5 of the chapter in which the reference occurs. Bibliographical references will be given as [10], referring to reference number 10 listed in the bibliography. We use the symbol ▮ to indicate the end of a proof or the end of a statement of a theorem if no proof is forthcoming.

The first-named author wishes to acknowledge partial support from the National Science Foundation during the preparation of this work.

Boulder, Colorado Albert T. Lundell

Lafayette, Indiana Stephen Weingram

Contents

CHAPTER 0

PRELIMINARIES

The purpose of this initial chapter is to clarify our peculiarities of notation and meaning. Our terminology from general topology will usually coincide with that of a standard text such as Kelley [22] with the following exceptions. A compact space is a space that has the finite subcover property *and is Hausdorff*. Thus a locally compact space will always be Hausdorff. In addition, a normal space is one in which disjoint closed sets can be separated by disjoint open neighborhoods *and is Hausdorff*. Thus a normal space for us is what Kelley would call T_4. This also applies to perfect normality.

Our other conventions of notation and terminology are as follows.

1. A *closed Euclidean n-cell* E^n is a homeomorphic image of the Euclidean n-cube I^n, the cartesian product of n copies of the closed unit interval $I = \{t \in \mathbf{R} \mid 0 \leq t \leq 1\}$. Note that the product of E^n and E^m is an $(n+m)$-cell. By a standard exercise, any closed bounded convex subset of a Euclidean space $\mathbf{R}^n$ that contains an interior point is homeomorphic to a closed Euclidean n-cell, and it is easy to see that this homeomorphism can be chosen so that it restricts to a homeomorphism of boundaries (see section 5). Other n-cells besides the n-cube with which we will be concerned are the unit n-disc $D^n = \{(t_1, t_2, \cdots, t_n) \in \mathbf{R}^n \mid \sum t_i^2 \leq 1\}$, and the *standard n-simplex* $\Delta^n = \{(t_0, t_1, \cdots, t_n) \in \mathbf{R}^{n+1} \mid t_i \geq 0 \text{ for } 0 \leq i \leq n \text{ and } \sum_i t_i = 1\}$. The *boundary* $\dot{E}^n$ of a closed Euclidean n-cell embedded as a convex set in $\mathbf{R}^n$ is its topological frontier and is

homeomorphic to the standard $(n-1)$-*sphere*

$$\mathbf{S}^{n-1} = \{ (t_1, t_2, \cdots, t_n) \in \mathbf{R}^n \mid \textstyle\sum_i t_i^2 = 1 \}.$$

An *open n-cell* is a homeomorphic image of $E^n - \dot{E}^n$. All these spaces except the open n-cell are compact, and the open n-cell is locally compact, in fact, the open n-cell is homeomorphic to Euclidean space $\mathbf{R}^n$.

2. If $\{X_\alpha \mid \alpha \in A\}$ is a collection of topological spaces, the *disjoint union* of this collection is a topological space, the underlying set of which is the disjoint union of the sets X_α and the topology of which can be described as follows. A set $U \subset \bigcup_\alpha X_\alpha$ is open if and only if each intersection $U \cap X_\alpha$ is open in the topology of X_α. If we are given a map $f_\alpha : X_\alpha \to Y$ for each $\alpha \in A$, the disjoint union of this family of maps is the map $f = \bigcup_\alpha f_\alpha : \bigcup_\alpha X_\alpha \to Y$ such that $f \mid X_\alpha = f_\alpha$. Clearly the disjoint union of a family of continuous maps is a continuous map. Note that for each $\beta \in A$, the inclusion $i_\beta : X_\beta \to \bigcup_\alpha X_\alpha$ is a continuous injection.

If a covering $\mathcal{B} = \{B_\alpha \mid \alpha \in A\}$ of a space X is such that a subset $U \subset X$ is open if and only if it meets each B_α in an open subset of B_α, we say that the covering $\mathcal{B}$ *dominates the space* X or that X has the *weak topology with respect to* $\mathcal{B}$. Clearly a disjoint union of spaces has the weak topology with respect to the family of summands.

3. If $f,g : X \to Y$ are continuous maps, a *homotopy* between f and g is a map $H : X \times I \to Y$ such that $H(x,0) = f(x)$ and $H(x,1) = g(x)$. If there is a homotopy between the maps $f,g : X \to Y$, we say that f and g are *homotopic* and write $f \simeq g$. It is easy to establish that homotopy defines an equivalence relation on the set of all maps from X to Y. The set of equivalence classes is usually denoted by $\pi(X;Y)$. If $f \simeq g$ are maps of X to Y, $h : Y \to Z$, and $k : W \to X$, then $hf \simeq hg$ and $fk \simeq gk$. Thus h induces a function $\pi(h) = h_\# : \pi(X;Y) \to \pi(X;Z)$, and k induces a function $\pi(k) = k^\# : \pi(X;Y) \to \pi(W;Y)$. One easily checks that $\pi(\ \ ;Y)$ and $\pi(X;\ \ )$ are contravariant and covariant functors, respectively, to the category of sets and functions.

If $A \subset X$ and $f,g:X \to Y$ are maps such that $f \mid A = g \mid A$, then f and g are *homotopic rel A* if there is a homotopy $H:X \times I \to Y$ between f and g such that $H(a,t) = f(a) = g(a)$ for all $a \in A$ and all $t \in I$.

Two spaces X and Y are said to have the same *homotopy type* if there exist maps $f:X \to Y$ and $g:Y \to X$ such that the compositions gf and fg are homotopic to the identity maps of X and Y, respectively. The maps are called *homotopy equivalences* and we say that one is the *homotopy inverse* of the other. Notice that if f has a left homotopy inverse g' and a right homotopy inverse g'', then either is a two-sided homotopy inverse, because $g''f \simeq g'fg''f \simeq g'f \simeq$ identity. An important special case of homotopy equivalence is the case of a deformation retract. A subspace $A \subset X$ is a *retract* of X if there is a continuous retraction map $r:X \to A$ such that $r \mid A$ is the identity map of A. A subspace $A \subset X$ is a *deformation retract* of X if there is a retraction $r:X \to A$ which is homotopic to the identity map of X via a homotopy which maps $A \times I$ into A. The subspace $A \subset X$ is a *strong deformation retract* of X if the homotopy of the retraction map with the identity of X is relative to the subspace A, i.e., if H is the homotopy, then $H(a,t) = r(a) = a$ for all $a \in A$ and $t \in I$. Finally, a space X *deforms into the subspace* A if there is a homotopy of the identity with a map $f:X \to A$; the homotopy is called a *deformation* of X into A. Note that a deformation retraction is the identity on A at $t=1$, but this need not be the case for a deformation.

We extend these concepts to *n-ads* which are n-tuples consisting of a space and $n-1$ subspaces. If $\boldsymbol{X} = (A_0 = X; A_1, \cdots, A_{n-1})$ (where $A_i \subset A_0 = X$ for $1 \le i \le n-1$) and $\boldsymbol{Y} = (B_0 = Y; B_1, \cdots, B_{n-1})$ are n-ads, then a map of n-ads $\boldsymbol{f}:\boldsymbol{X} \to \boldsymbol{Y}$ is a map $f:X \to Y$ such that $f(A_i) \subset B_i$ for $1 \le i \le n-1$. The product of a space C with the n-ad $\boldsymbol{X}$ is the n-ad $\boldsymbol{X} \times C = (X \times C; A_1 \times C, A_2 \times C, \cdots, A_{n-1} \times C)$, and an n-ad homotopy is a map $\boldsymbol{H}:\boldsymbol{X} \times I \to \boldsymbol{Y}$ of n-ads. A retraction of n-ads is a retraction that restricts to a retraction on each A_i, and deformation retraction and deformation are defined analogously. We denote the homotopy classes of maps of the n-ad $\boldsymbol{X}$ into the n-ad $\boldsymbol{Y}$ by $\pi(\boldsymbol{X};\boldsymbol{Y})$ or by $\pi(X_0, \cdots, X_{n-1}; Y_0, \cdots, Y_{n-1})$.

In particular, let X be a space and let $x \in X$. We write $\boldsymbol{X} = (X,x)$ and $\boldsymbol{S}^n = (E^n, \dot{E}^n)$. Then $\pi(\boldsymbol{S}^n;\boldsymbol{X}) = \pi(E^n, \dot{E}^n; X, x)$ is usually denoted by $\pi_n(X,x)$, or by $\pi_n(X)$ if X is pathwise connected. If

$x \in A \subset X$, and $e \in \dot{E}^n$, then $\pi(E^n,\dot{E}^n,e;X,A,x)$ is denoted by $\pi_n(X,A,x)$, or by $\pi_n(X,A)$ if A is pathwise connected. If X is pathwise connected and $\pi_1(X)$ is a single element, X is said to be *simply connected.*

These sets are usually studied in an algebraic setting where it is proved that $\pi_n(X,x)$ and $\pi_{n+1}(X,A,x)$ have group structures when $n \geq 1$, and the induced maps $f_\#$ are homomorphisms of this group structure. With this structure, the set $\pi_n(X)$ is called the *n*th *homotopy group* of the (pathwise connected) space X, and $\pi_n(X,A,x)$ the *n*th *relative homotopy group* of the pair (X,A) at x.

The proofs of the following propositions are omitted; the interested reader may regard them as exercises or may refer to Spanier [35, Chapter I].

PROPOSITION. *The set $\pi_n(X,x)$ reduces to a single element if and only if any map of $\dot{E}^n$ into the path component C of x extends to a map of E^n into C. The set $\pi_n(X,A,x)$ reduces to a single element if and only if any map of the pair $(E^n,\dot{E}^n)$ into (X,A) is homotopic relative to A to a map of E^n into A.*

PROPOSITION. *If $i:A \to X$ is an inclusion map and X and A are path connected spaces, then the induced maps $i_\#:\pi_n(A) \to \pi_n(X)$ are all bijective if and only if each set $\pi_n(X,A)$ reduces to a single element.*

PROPOSITION. *If X and Y are pathwise connected spaces and $f:X \to Y$ is a map, then the induced map $f_\#:\pi_n(X) \to \pi_n(Y)$ is an injection if and only if it sends no element onto the class of the constant map except the class of the constant map.*

4. A pair (X, A) has the *homotopy extension property* with respect to a space Y if for every map $f:X \to Y$ and homotopy $h:A \times I \to Y$ such that $h(a,0) = f(a)$, there is a homotopy $H:X \times I \to Y$ such that $H(x,0) = f(x)$ and $H(a,t) = h(a,t)$ for $(a,t) \in A \times I$. If (X,A) has the homotopy extension property with respect to all spaces Y, the inclusion map $i:A \to X$ is called a *cofibration.*

5. A topological space X is a *topological n-manifold* provided each point $x \in X$ has a neighborhood homeomorphic to Euclidean n-space $\mathbf{R}^n$.

Standard examples of topological n-manifolds are $\mathbf{R}^n$ and the n-sphere $\mathbf{S}^n$. As a neighborhood of $x \in \mathbf{S}^n$ which is homeomorphic to $\mathbf{R}^n$ we may take $\mathbf{S}^n - \{-x\}$, with the homeomorphism given by stereographic projection from the point $-x$.

We will need the following theorem and its corollaries, proofs of which may be found in Eilenberg and Steenrod [10, p. 303]. We regard it as nonelementary.

*THEOREM. (*Invariance of domain*). *Let A be a subset of the topological n-manifold X, let B be a subset of the topological n-manifold Y, and let $f: A \to B$ be a homeomorphism. If A is open in X, then B is open in Y.*

*COROLLARY. *Let $m < n$, and let A be a nonempty open subset of the topological n-manifold X. Then A is not homeomorphic to any open subset of a topological m-manifold Y.*

*COROLLARY. *If φ is any homeomorphism of n-cells, then φ maps interior points to interior points and boundary points to boundary points.*

* An asterisk on a passage indicates that a knowledge of general topology or the simplest elements of homology are not sufficient.

CHAPTER I

COMBINATORIAL CELL COMPLEXES

In this chapter we state the basic definitions and propositions about combinatorial cell complexes and work out some examples in detail. We study certain maps of cell complexes that preserve "enough" of the combinatorial structure and construct the product, quotient, and adjunction complexes.

1. DEFINITIONS

Definition **1.1.** Let X be a set. A *cell structure* on X is a pair (X,Φ), where Φ is a collection of maps of closed Euclidean cells into X satisfying the following conditions.

(i) If $\varphi\in\Phi$ and φ has domain E^n, then φ is injective on $E^n-\dot{E}^n$.
(ii) The images $\{\varphi(E^n-\dot{E}^n)\mid\varphi\in\Phi\}$ partition X, i.e., they are disjoint and have union X.
(iii) If $\varphi\in\Phi$ has domain E^n, then $\varphi(\dot{E}^n)\subset\bigcup\{\psi(E^k-\dot{E}^k)\mid\psi\in\Phi$ has domain E^k and $k\leq n-1\}$.

If $\varphi\in\Phi$ and φ has domain E^n, the image set $\varphi(E^n)=\sigma^n$ is called an *n-cell* or *closed n-cell* of (X,Φ), and we say φ is a *characteristic map* for the cell σ^n. Thus Φ is a set of characteristic maps for the cells of (X,Φ). Also $\varphi(\dot{E}^n)=\dot{\sigma}^n$ is called the *boundary* of the cell σ^n, and $\varphi(E^n-\dot{E}^n)$ is called its *interior*. If $n>0$, $\varphi(E^n-\dot{E}^n)$ is called an *open n-cell.* This terminology is classical, but much of it is unfortunate. For example, in the absence of a topology on X, the closed and open cells are neither open nor closed. Moreover, under the usual topologies one puts on a set with a cell structure, open cells need not be open sets and closed cells need not be (homeomorphs of) cells.

The union $\bigcup\{\psi(E^k-\dot{E}^k)\mid \psi\in\Phi \text{ has domain } E^k \text{ and } k\leq n-1\}=X^{n-1}$, which appears in condition (iii), is called the $(n-1)$-*skeleton* of the cell structure. Thus, for each n, $\dot{\sigma}^n=\varphi(\dot{E}^n)\subset X^{n-1}$.

The following three remarks round out the definition given above.

LEMMA **1.2.** *Let (X,Φ) be a cell structure. Then*

(1) *if σ^n is a cell with characteristic map φ, then $\varphi(E^n-\dot{E}^n)=\sigma^n-\dot{\sigma}^n$ is the interior of σ^n;*

(2) *each n-cell is a subset of X^n;*

(3) *for each n, $X^n=\bigcup\{\sigma^k\mid \sigma^k$ is a k-cell of (X,Φ) and $k\leq n\}$.*

Proof. (1) By condition (iii), $\dot{\sigma}^n$ is contained in X^{n-1}, which is disjoint from $\varphi(E^n-\dot{E}^n)$, since the sets $\psi(E^q-\dot{E}^q)$ partition X. Hence $\varphi(E^n-\dot{E}^n)\cup\varphi(\dot{E}^n)$ is a disjoint union equal to σ^n, and therefore $\sigma^n-\dot{\sigma}^n=\varphi(E^n-\dot{E}^n)$.

(2) By hypothesis $\sigma^n-\dot{\sigma}^n$ lies in X^n and by condition (iii), $\dot{\sigma}^n$ lies in $X^{n-1}\subset X^n$. Therefore $\sigma^n=(\sigma^n-\dot{\sigma}^n)\cup\dot{\sigma}^n\subset X^n$.

(3) Since each $\sigma^n-\dot{\sigma}^n\subset\sigma^n$, the n-skeleton X^n is certainly contained in the union of the cells of dimension $\leq n$. By part (2) above, the reverse inclusion is also valid. ▮[1]

We say that two cell structures (X,Φ) and (X,Φ') are *strictly equivalent* if there is a one-to-one correspondence between Φ and Φ' such that a characteristic function with domain E^n corresponds to a characteristic function with domain E^n, and corresponding functions differ only by a reparametrization of their domain. That is, if φ and φ' are corresponding functions of Φ and Φ', respectively, then $\varphi'=\varphi\cdot h$, where $h\colon(E^n,\dot{E}^n)\to(E^n,\dot{E}^n)$ is a homeomorphism of pairs. We leave it to the reader to check that this is an equivalence relation on the collection of cell structures on the set X. If (X,Φ) is a cell structure, let $\mathcal{S}_\Phi$ consist of all pairs $(\sigma^n,[\varphi])$, where $\sigma^n=\varphi(E^n)$ and $[\varphi]$ is the strict equivalence class of $\varphi\in\Phi$. For convenience, we will denote such a pair by σ^n or $\varphi(E^n)$ with the class of φ understood. Note that if (X,Φ) and (X,Φ') are strictly equivalent cell structures, then $\mathcal{S}_\Phi=\mathcal{S}_{\Phi'}$.

[1] The symbol ▮ indicates the end of a proof or the end of a statement of a theorem if no proof is given.

Definition **1.3.** A *cell complex* on a set X or a *cellular decomposition* of a set X is an equivalence class of cell structures (X,Φ) under the equivalence relation of strict equivalence. A cell complex on X will be denoted by a pair $(X,\mathcal{S})$, where $\mathcal{S}=\mathcal{S}_\Phi$ for some representative cell structure (X,Φ). The set $\mathcal{S}$ is called the set of (closed) cells of $(X,\mathcal{S})$.

In order to simplify some of the notation, when the dimension of a member of $\mathcal{S}$ is not important, we will write $\sigma=\sigma^n$. At times we index the characteristic maps by their image, and if dimension is not important, we index the domain of a characteristic map by its image. Thus if $\sigma\in\mathcal{S}$, it has a characteristic map φ_σ and $\varphi_\sigma:E_\sigma\to X$, where E_σ is a Euclidean cell.

Definition **1.4.** A *subcomplex* $(A,\mathfrak{I})$ of a cell complex $(X,\mathcal{S})$, which we denote by $(A,\mathfrak{I})\subset(X,\mathcal{S})$, is a cell complex such that $A\subset X$ and $\mathfrak{I}\subset\mathcal{S}$. We adopt the convention that the empty set together with the empty set of cells is a complex, and if $(X,\mathcal{S})$ is a cell complex, we write $(\varnothing,\varnothing)=(X^{-1},\mathcal{S}^{-1})$ for its empty subcomplex.

As an example, if $(X,\mathcal{S})$ is any cell complex, let $\mathcal{S}^n=\{\sigma^p\in\mathcal{S}\mid p\leq n\}$. Then $(X^n,\mathcal{S}^n)$ is a subcomplex of $(X,\mathcal{S})$ called the *n-skeleton* of $(X,\mathcal{S})$. Note that

$$(X^{-1},\mathcal{S}^{-1})\subset(X^0,\mathcal{S}^0)\subset(X^1,\mathcal{S}^1)\subset\cdots\subset(X,\mathcal{S}).$$

LEMMA **1.5.** *Let $(X,\mathcal{S})$ be a cell complex, and let $\{(X_\gamma,\mathcal{S}_\gamma)\mid\gamma\in\Gamma\}$ be a family of subcomplexes of $(X,\mathcal{S})$. Then $(\bigcup_\gamma X_\gamma,\bigcup_\gamma\mathcal{S}_\gamma)$ and $(\bigcap_\gamma X_\gamma,\bigcap_\gamma\mathcal{S}_\gamma)$ are subcomplexes of $(X,\mathcal{S})$.*

The proof is left as an exercise. ∎

The following proposition characterizes the cells of a subcomplex.

PROPOSITION **1.6.** *Let $(A,\mathfrak{I})$ be a subcomplex of the cell complex $(X,\mathcal{S})$, and let σ be a cell in $\mathcal{S}$. Then σ is a cell of $\mathfrak{I}$ if and only if $(\sigma-\dot{\sigma})\cap A\neq\varnothing$.*

Proof. Clearly, if $\sigma \in \mathfrak{I}$, then $\sigma \subset A$, so that $(\sigma - \dot{\sigma}) \cap A \neq \varnothing$. Suppose that $(\sigma - \dot{\sigma}) \cap A \neq \varnothing$. Choose characteristic maps Φ for $(X,\mathcal{S})$ and Ψ for $(A,\mathfrak{I})$ such that $\Psi \subset \Phi$. If $y \in (\sigma - \dot{\sigma}) \cap A$, then $y \in \varphi_\sigma(E_\sigma - \dot{E}_\sigma)$ for $\varphi_\sigma \in \Phi$. Since the open cells partition X, in order that $y \in A$, we must have $\varphi_\sigma \in \Psi$. Thus $\sigma \in \mathfrak{I}$. ∎

Definitions **1.7.** A cell complex $(X,\mathcal{S})$ is *finite* or *countable* if $\mathcal{S}$ is a finite or countable set. It is *locally finite* or *locally countable* if each closed cell meets only a finite or countable number of cells, and *closure finite* if each n-cell meets only a finite number of open cells $\sigma^p - \dot{\sigma}^p$ with $p < n$.

A cell is called a *regular cell* if it has one (and hence every equivalent) characteristic map bijective on all E^n. A cell complex is *regular* if all its cells are regular. A cell complex is *normal* if, for each cell σ, the subset σ carries the structure of a subcomplex We will broaden the definition of regular cell complex later in Chapter III.

The cell complex $(X,\mathcal{S})$ has *dimension n* if it has no cells σ^p of dimension p greater than n, and at least one cell of dimension n. If a cell complex is not of dimension n for any n, we say that it is *infinite dimensional*. The empty complex has dimension -1.

2. EXAMPLES

Although we have not yet defined a topology on a cell complex, most of the sets for which we give a cellular decomposition are well-known topological spaces, and the characteristic maps we give will be continuous with respect to these topologies. In fact, although we do not give the proofs, the restrictions of characteristic maps $\varphi: E^n - \dot{E}^n \to \sigma^n - \dot{\sigma}^n$ will be homeomorphisms, and (hence) in the case of regular cells, homeomorphisms on all of E^n. As domains of characteristic functions we will take the unit n-disc, except in example 2.7.

Example **2.1.** *The n-sphere and the* $(n+1)$*-disc.* For any $n \geq 0$, the n-sphere

$$\mathbf{S}^n = \{x = (x_0, x_1, \cdots, x_n) \in \mathbf{R}^{n+1} \mid \langle x,x \rangle = x_0^2 + x_1^2 + \cdots + x_n^2 = 1\}$$

has a cell structure $(\mathbf{S}^n, \Phi)$, where Φ consists of two functions, $\varphi^0 : D^0 \to \mathbf{S}^n$ and $\varphi^n : D^n \to \mathbf{S}^n$. We define

$$\varphi^0(x) = (1,0, \cdots, 0)$$

$$\varphi^n(x) = (2\langle x,x\rangle - 1, 2x_1\sqrt{1-\langle x,x\rangle}, \cdots, 2x_n\sqrt{1-\langle x,x\rangle}).$$

Note that if $\mathcal{S} = \{\sigma^0, \sigma^n\}$ is the collection of cells, $\sigma^0 = (1,0, \cdots, 0)$ and $\sigma^n = \mathbf{S}^n$. Also observe that the set of the k-skeleton of $\mathbf{S}^n$ in this decomposition is the 0-skeleton for $k < n$ and is all of $\mathbf{S}^n$ for $k \geq n$.

To get a cell structure on the $(n+1)$-disc, we take the two characteristic maps above together with the identity map $\varphi^{n+1} : D^{n+1} \to D^{n+1}$. This gives us a cell complex $(D^{n+1}, \mathcal{D}^{n+1})$ which has exactly three cells and contains the cell complex $(\mathbf{S}^n, \mathcal{S})$ given above as a subcomplex. Note that the cell of dimension $n+1$ is a regular cell.

Example **2.2.** *The n-sphere and the* $(n+1)$*-disc, regular cell structures.* We give a cell structure $(\mathbf{S}^n, \Phi^n)$, where

$$\Phi^n = \{\varphi_+{}^0, \varphi_-{}^0, \varphi_+{}^1, \cdots, \varphi_+{}^n, \varphi_-{}^n\}$$

and each $\varphi_\pm{}^k : D^k \to \mathbf{S}^n$ is a bijection. We define

$$\varphi_+{}^k(x) = (\sqrt{1-\langle x,x\rangle}, x_1, x_2, \cdots, x_k, 0, \cdots, 0)$$

$$\varphi_-{}^k(x) = (-\sqrt{1-\langle x,x\rangle}, x_1, x_2, \cdots, x_k, 0, \cdots, 0),$$

for $k = 0,1, \cdots, n$. We point out that if

$$\mathcal{S}^n = \{\sigma_+{}^0, \sigma_-{}^0, \sigma_+{}^1, \cdots, \sigma_+{}^n, \sigma_-{}^n\}$$

is the set of cells, $(\mathbf{S}^k, \mathcal{S}^k)$ is a subcomplex of $(\mathbf{S}^n, \mathcal{S}^n)$, in fact, $(\mathbf{S}^k, \mathcal{S}^k)$ is the k-skeleton of $(\mathbf{S}^n, \mathcal{S}^n)$. Another useful property of this decomposition is that if $T : \mathbf{S}^n \to \mathbf{S}^n$ is the antipodal map $T(x) = -x$, we have $T(\sigma_+{}^k) = \sigma_-{}^k$ and $T(\sigma_-{}^k) = \sigma_+{}^k$ for $k = 0,1, \cdots, n$.

To get a decomposition of the $(n+1)$-disc we again use the identity map $\varphi^{n+1}:D^{n+1}\to D^{n+1}$ as the characteristic map for the cell of dimension $(n+1)$, and the characteristic maps for the decomposition of the n-sphere for the lower dimensional ones. Thus we have $\mathbf{S}^n$ as a (regular) subcomplex of a regular cell structure on D^{n+1}.

Example **2.3.** *A cartesian product of spheres.* Let $X=\mathbf{S}^m\times\mathbf{S}^n$, let $p_0=(1,0,\cdots,0)\in\mathbf{S}^m$, and let $q_0=(1,0,\cdots,0)\in\mathbf{S}^n$. We define a cell structure (X,Ψ), where $\Psi=\{\psi^0,\psi^m,\psi^n,\psi^{m+n}\}$ and $\psi^k:D^k\to X$. Let $h_{m,n}:(D^{m+n},\dot{D}^{m+n})\to(D^m\times D^n,D^m\times\dot{D}^n\cup\dot{D}^m\times D^n)$ be a homeomorphism of pairs. We define

$$\psi^0(x)=(p_0,q_0)$$

$$\psi^m(x)=(\varphi^m(x),q_0)$$

$$\psi^n(x)=(p_0,\varphi^n(x))$$

$$\psi^{m+n}(x)=(\varphi^m\times\varphi^n)h_{m,n}(x),$$

where φ^m and φ^n are the characteristic maps of example 2.1.

Example **2.4.** *The projective spaces.* Let **F** be one of the topological fields **R** (real numbers), **C** (complex numbers), or **H** (quaternions), and let d_F denote the dimension of **F** as an algebra over **R**, so that $d=d_F=1$, 2, or 4. Let $\mathbf{F}^n$ denote the right vector space of n-tuples over **F** with the usual inner product $\langle x,y\rangle=\sum_i x_i\bar{y}_i$, where $x=(x_1,x_2,\cdots,x_n)$ and $y=(y_1,y_2,\cdots,y_n)$, and $\bar{u}$ denotes the conjugate of $u\in\mathbf{F}$. Let $\mathbf{G}_F=\{u\in F\mid u\bar{u}=1\}$, which is easily seen to be a topological group. Clearly $\mathbf{G}_F$ is homeomorphic to $\mathbf{S}^{d-1}$ as a topological space. Also note that

$$\mathbf{S}^{dn-1}=\{x\in\mathbf{F}^n\mid \langle x,x\rangle=1\},$$

and that scalar multiplication

$$(x_1,x_2,\cdots,x_n)u=(x_1u,x_2u,\cdots,x_nu),\qquad u\in\mathbf{G}_F,$$

preserves the sphere $\mathbf{S}^{dn-1}$, i.e., the group $\mathbf{G}_F$ operates on $\mathbf{S}^{dn-1}$. One defines an equivalence relation on $\mathbf{S}^{dn-1}$ via this operation of $\mathbf{G}_F$, i.e., $x,y \in \mathbf{S}^{dn-1}$ are equivalent provided there is an element $u \in \mathbf{G}_F$ such that $y=xu$. The quotient space under this relation is called real, complex, or quaternionic projective space according as the field $\mathbf{F}$ is $\mathbf{R}$, $\mathbf{C}$, or $\mathbf{H}$. It will be denoted by $\mathbf{FP}^{n-1}$, and we let $\pi_F: \mathbf{S}^{dn-1} \to \mathbf{FP}^{n-1}$ denote the quotient map and

$$\pi_F(x_1,x_2, \cdots, x_n) = [x_1,x_2, \cdots, x_n].$$

The (nonunique) n-tuple $(x_1,x_2, \cdots, x_n)$ is called a set of projective coordinates of the point $[x_1,x_2, \cdots, x_n]$. Two n-tuples $(x_1,x_2, \cdots, x_n)$, $(y_1,y_2, \cdots, y_n)$ determine the same point of $\mathbf{FP}^{n-1}$ if and only if there is a $u \in \mathbf{S}^{d-1}$ such that $x_k u = y_k$ for $k=1,2, \cdots, n$.

The inclusion map $\mathbf{F}^n \subset \mathbf{F}^{n+1}$ defined by $(x_1,x_2, \cdots, x_n) \to (x_1,x_2, \cdots, x_n,0)$ induces inclusions $\mathbf{S}^{dn-1} \to \mathbf{S}^{d(n+1)-1}$ and $\mathbf{FP}^{n-1} \to \mathbf{FP}^n$. Finally, one defines $\mathbf{FP}^\infty = \mathbf{U}_n \mathbf{FP}^n$ with a suitable topology and calls it infinite dimensional real, complex, or quaternionic projective space according as $\mathbf{F} = \mathbf{R}$, $\mathbf{C}$, or $\mathbf{H}$.

We define a cellular structure $(\mathbf{FP}^{n-1}, \Phi^{n-1})$ on the set $\mathbf{FP}^{n-1}$ as follows. Take $\Phi_F = (\varphi_F{}^0, \varphi_F{}^d, \cdots, \varphi_F{}^{d(n-1)}\}$, where $\varphi_F{}^{kd}: D^{kd} \to \mathbf{FP}^{n-1}$ is defined by

$$\varphi_F{}^{kd}(x_1,x_2, \cdots, x_k) = [x_1,x_2, \cdots, x_k, \sqrt{1-\langle x,x\rangle},0, \cdots, 0],$$

and we have identified D^{kd} with $\{x \in \mathbf{F}^k \mid \langle x,x\rangle \leq 1\}$. One easily checks that $(\mathbf{FP}^{n-1}, \Phi_F{}^{n-1})$ is a cell structure. Moreover, if $\mathcal{S}^{n-1}$ is the set of cells, $(\mathbf{FP}^{n-1}, \mathcal{S}^{n-1})$ is a subcomplex of $(\mathbf{FP}^n, \mathcal{S}^n)$. If we set $\mathcal{S} = \mathbf{U}_n \mathcal{S}^n$, $(\mathbf{FP}^\infty, \mathcal{S})$ is a cell complex with n-skeleton $(\mathbf{FP}^n, \mathcal{S}^n)$.

It is worthwhile making a few remarks about the case $\mathbf{F} = \mathbf{R}$. Note that in this case $\mathbf{G}_F = \mathbf{G}_R = \{1,-1\}$, and the operation of $\mathbf{G}_R$ on $\mathbf{S}^{n-1}$ is either the identity operation or the antipodal map. If we take the cell complex on $\mathbf{S}^{n-1}$ given in example 2.2, the projection $\pi_R: \mathbf{S}^{n-1} \to \mathbf{RP}^{n-1}$ identifies the cells $\sigma_+{}^k$ and $\sigma_-{}^k$. The map π_R

is a local homeomorphism and a two-sheeted covering map, i.e., $\pi_R^{-1}(z)$ consists of two elements of S^{n-1}.

Example **2.5.** *Grassman varieties.* We can generalize the example of the projective spaces $\mathbf{FP}^n$. The Grassman variety $\mathbf{FG}(n,m)$ is the set of n-planes through the origin of the vector space $\mathbf{F}^{n+m}$, where $\mathbf{F}$ is one of the fields $\mathbf{R}$, $\mathbf{C}$, or $\mathbf{H}$. We will describe the cellular structure of $\mathbf{FG}(n,m)$ by identifying with a certain quotient set. Let D be the subset of $(\mathbf{F}^{n+m})^n$ consisting of n-tuples of vectors which are linearly dependent. Define an equivalence relation $\mathfrak{R}$ on $(\mathbf{F}^{n+m})^n - D$ by defining two such n-tuples of vectors equivalent if they determine the same n-plane in $\mathbf{F}^{n+m}$. Then the quotient set $[(\mathbf{F}^{n+m})^n - D]/\mathfrak{R}$ is $\mathbf{FG}(n,m)$. Another description which is more useful is given as follows. Let $\mathbf{FQ}(n,m)$ be the set of $n \times (n+m)$ matrices over $\mathbf{F}$ which are in reduced echelon form with each row vector normalized to length 1. Such a matrix will have the following form:

$$\begin{bmatrix} \begin{matrix} n \times p_n \end{matrix} & \begin{matrix} r_1 & 0 \\ 0 & \\ 0 & (n-1)\times p_{n-1} \\ \vdots & \\ 0 & \end{matrix} & \begin{matrix} 0 \\ r_2 & 0 \\ 0 & (n-2)\times p_{n-2} \\ \vdots & \\ 0 & \end{matrix} & \begin{matrix} 0 \\ 0 \\ r_3 \\ \vdots \\ 0 \cdots 0 \boxed{1 \times p_1} \; r_n \; 0 \cdots 0 \end{matrix} & \begin{matrix} 0 \\ 0 \\ 0 \\ \vdots \\ \end{matrix} \end{bmatrix}$$

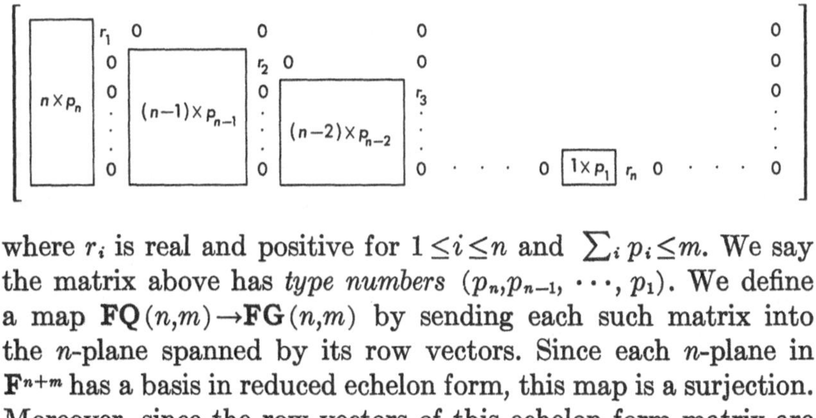

where r_i is real and positive for $1 \le i \le n$ and $\sum_i p_i \le m$. We say the matrix above has *type numbers* $(p_n, p_{n-1}, \cdots, p_1)$. We define a map $\mathbf{FQ}(n,m) \to \mathbf{FG}(n,m)$ by sending each such matrix into the n-plane spanned by its row vectors. Since each n-plane in $\mathbf{F}^{n+m}$ has a basis in reduced echelon form, this map is a surjection. Moreover, since the row vectors of this echelon form matrix are of norm 1 and have positive last coefficient, this map is injective and hence a bijection.

We will define a cellular decomposition of $\mathbf{FG}(n,m)$ which will have one cell for each sequence $(p_n, p_{n-1}, \cdots, p_1)$, where $p_i \ge 0$ for $1 \le i \le n$ and $\sum_i p_i \le m$. The dimension of the cell corresponding to the sequence $(p_n, p_{n-1}, \cdots, p_1)$ will be $d_F(\sum_k k p_k)$, where d_F is the dimension of $\mathbf{F}$ as an algebra over $\mathbf{R}$.

Regard the disc D^{qd} as the unit disc in $\mathbf{F}^q$. If $N=d_F(\sum_k kp_k)$, write $E^N=D^{p_nd}\times D^{(p_n+p_{n-1})d}\times\cdots\times D^{(p_n+\cdots+p_1)d}$. Then

$$\varphi_{(p_n,p_{n-1},\cdots,p_1)}:E^N\to\mathbf{FG}(n,m)$$

will be the map $\mathbf{FQ}(n,m)\to\mathbf{FG}(n,m)$ composed with the map $E^N\to\mathbf{FQ}(n,m)$ which takes the n-tuple of vectors $(v^1,v^2,\cdots,v^n)$ to the matrix

$$\begin{bmatrix} v_1^1\cdots v_{p_n}^1 & (1-|v^1|^2)^{1/2} & 0\cdots 0 & 0 & \cdots & 0 & 0\cdots 0 \\ v_1^2\cdots v_{p_n}^2 & 0 & v_{p_n+1}^2\cdots v_{p_n+p_{n-1}}^2 & (1-|v^2|^2)^{1/2} & \cdots & \vdots & \vdots \\ \vdots & \vdots & \vdots & \vdots & & \vdots & \vdots \\ v_1^n\cdots v_{p_n}^n & 0 & v_{p_n+1}^n\cdots v_{p+p_{n-1}}^n & 0 & \cdots & (1-|v^n|^2)^{1/2} & 0\cdots 0 \end{bmatrix}$$

Observe that the boundary of E^N consists of points for which some vector v^k lies in the boundary $\dot{D}^{(p_n+\cdots+p_{n-k+1})d}$, i.e., has length 1 so that, in the kth row, the radical is zero. An inductive proof establishes that this matrix can be reduced to echelon form (without changing the subspace generated by the row vectors), and therefore the new type numbers are

$$(p_n',p_{n-1}',\cdots,p_1') \qquad \text{and} \qquad \textstyle\sum_k kp_k'<\sum_k kp_k.$$

Thus $\varphi_{(p_n,p_{n-1},\cdots,p_1)}(\dot{E}^N)$ is contained in a union of images of cells of lower dimension. One easily checks that with these characteristic maps we have defined a cellular decomposition of $\mathbf{FG}(n,m)$.

A few remarks should be made about this cellular decomposition. First observe that $\mathbf{FG}(n,m)$ has one 0-cell and one d_F-cell which correspond to the matrices

$$\begin{bmatrix} 1 & 0\cdots 0 & 0\cdots 0 \\ 0 & 1\cdots 0 & 0\cdots 0 \\ \vdots & & \vdots \\ 0 & 0\cdots 1 & 0\cdots 0 \end{bmatrix}$$

$$\begin{bmatrix} 1 & 0\cdots 0 & 0 & 0 & 0\cdots 0 \\ 0 & 1\cdots 0 & 0 & 0 & 0\cdots 0 \\ \vdots & & & & \vdots \\ 0 & 0\cdots 1 & 0 & 0 & 0 \quad 0 \\ 0 & 0\cdots 0 & x & (1-|x|^2)^{1/2} & 0\cdots 0 \end{bmatrix},$$

respectively. In general, $\mathbf{FG}(n,m)$ has cells only of dimension kd_F, and the number of cells of dimension kd_F is the number $\rho_k(n,m)$ of sequences $(p_n, \cdots, p_1)$ with $p_i>0$ and $\sum_i ip_i=k\leq m$.

Another point we should mention is that $\mathbf{FP}^n=\mathbf{FG}(1,n)$, and the decomposition given here coincides with that of example 2.4. Thus we have a true generalization of example 2.4.

Finally, note that the natural inclusions $\mathbf{F}^q\subset\mathbf{F}^{q+1}\subset\cdots$ induce inclusion maps $\mathbf{FG}(n,m)\subset\mathbf{FG}(n,m+1)\subset\cdots$ and the inclusions are as subcomplexes. Define $\mathbf{FG}_n=\mathbf{U}_{m>1}\mathbf{FG}(n,m)$. Then $\mathbf{FG}_n$ is a cell complex which, with a suitable topology, is called the *classifying space for* $\mathbf{F}$ *n-plane bundles.*

Example **2.6.** *Lens spaces.* Let p,q be relatively prime positive integers. We define a space $\mathbf{L}^n(p,q)$ called the $(2n-1)$-dimensional *lens space of type* p,q as in DeRahm [3].

Let $\mathcal{R}$ be the equivalence relation on

$$\mathbf{S}^{2n-1}=\{(z_1, \cdots, z_n)\in\mathbf{C}^n \mid \textstyle\sum_i |z_i|^2=1\}$$

defined by $(z_1, \cdots, z_n)\sim(\zeta z_1, \cdots, \zeta z_{n-1},\zeta^q z_n)$, where ζ is a primitive pth root of unity. Let $\mathbf{L}^n(p,q)$ be the quotient set, and $\eta:\mathbf{S}^{2n-1}\to\mathbf{L}^n(p,q)$ the quotient map. If $T:\mathbf{S}^{2n-1}\to\mathbf{S}^{2n-1}$ is the map defined by $T(z_1, \cdots, z_n)=(\zeta z_1, \cdots, \zeta z_{n-1},\zeta^q z_n)$, then T is a rotation of $\mathbf{S}^{2n-1}$, in fact, with characteristic numbers ζ,ζ^{-1} $(n-1)$-times repeated, and ζ^q,ζ^{-q}. For any point $z\in\mathbf{S}^{2n-1}$, the points $z,T(z), \cdots, T^{p-1}(z)$ are distinct and identified by the quotient map η. Also $T^p(z)=z$.

Let $(\mathbf{S}^{2n-1},\mathcal{S})$ be the following regular cell complex on $\mathbf{S}^{2n-1}$. For each k, $0\leq k\leq p-1$, and $r\leq n$, we have one $(2r-2)$-cell $\sigma_k^{2r-2}=\{(z_1, \cdots, z_r, 0, \cdots, 0)\in\mathbf{S}^{2n-1}\mid z_r=0 \text{ or } \operatorname{Arg} z_r=2\pi k/p\}$, and one $(2r-1)$-cell $\sigma^{2r-1}=\{(z_1, \cdots, z_r,0, \cdots, 0)\in\mathbf{S}^{2n-1}\mid z_r=0$ or $2\pi k/p\leq \operatorname{Arg} z_r\leq 2\pi(k+1)/p\}$. In order to simplify subsequent formulas, we let $D_C^{2s}=\{z=(z_1, \cdots, z_s)\in\mathbf{C}^s\mid \sum_i |z_i|^2\leq 1\}$ and define a characteristic map $\varphi_k^{2r-2}:D_C^{2r-2}\to\sigma_k^{2r-2}\subset\mathbf{S}^{2n-1}$ by the formula $\varphi_k^{2r-2}(z)=(z_1, \cdots, z_{r-1},(1-|z|^2)^{1/2}e^{2\pi ik/p},0, \cdots, 0)$. We define a characteristic map $\varphi_k^{2r-1}:D_C^{2r-2}\times D^1\to\sigma^{2r-1}\subset\mathbf{S}^{2n-1}$ as follows. First define a map $\varphi_k'^{2r-1}:D_C^{2r-2}\times D^1\to\sigma^{2r-1}\subset\mathbf{S}^{2n-1}$ by the formula $\varphi_k'^{2r-1}(z,x)=(z_1, \cdots, z_{r-1},(1-|z|)^{1/2}e^{2\pi i(x+2k+1)/p},0, \cdots, 0)$. The map $\varphi_k'^{2r-1}$ is a bijection on the interior of $D_C^{2r-2}\times D^1$ and is a surjection onto $\sigma_k'^{2r-1}$, but is not a bijection on all of $D_C^{2r-2}\times D^1$. There is, however, a continuous map h_s of $D_C^{2s}\times D^1$ onto itself such that $h_s(z,x)=(z,-1)$ for $|z|=1$, h_s maps $D_C^{2s}\times \dot{D}^1$ onto $(D_C^{2s}\times D^1)^{\cdot}$, and maps the interior of $D_C^{2s}\times D^1$ homeomorphically. It is clear that $\varphi_k'^{2r-1}=\varphi_k^{2r-1}h_{r-1}$ for a suitable bijection φ_k^{2r-1} of $D_C^{2r-2}\times D^1$ onto σ_k^{2r-1}. The maps $\varphi_k{}^r$ provide a regular decomposition of $\mathbf{S}^{2n-1}$. Note that $\dot{\sigma}_k^{2r}=\bigcup_j\sigma_j^{2r-1}$ and that $\dot{\sigma}_k^{2r-1}=\sigma_k^{2r-2}\cup\sigma_{k+1}^{2r-2}$ for $r\geq 1$.

Now observe that for each k, $T:\sigma_k{}^m\to\sigma_{k+1}^m$ is a bijection (where we make the identification $\sigma_p{}^m=\sigma_0{}^m$). Thus under the quotient map η, all cells of a given dimension will be identified. Let $(\mathbf{L}^n(p,q),\mathfrak{J})$ be the cell structure on $\mathbf{L}^n(p,q)$ given by one cell $\tau^m=\eta(\sigma_0{}^m)$ for each m, $0\leq m\leq 2n-1$, with characteristic map $\psi^m=\eta\varphi_0{}^m$. We point out that the lens space $\mathbf{L}^n(2,q)$ is just the projective space $\mathbf{RP}^{2n-1}$, and the cellular decomposition we have defined is just the decomposition of example 2.4.

The embedding $\mathbf{C}^n\to\mathbf{C}^{n+1}$ defined by $(z_1,\cdots,z_n)\to(z_1,\cdots,z_n,0)$ induces an embedding $\mathbf{L}^n(p,1)\to\mathbf{L}^{n+1}(p,1)$ as the subcomplex of the $(2n-1)$-skeleton. Thus we may define a cell complex on $\mathbf{L}^\infty(p,1)=\bigcup_n\mathbf{L}^n(p,1)$. The set $\mathbf{L}^\infty(p,1)$ has one cell in each dimension, and with a suitable topology is called a space of type $(Z_p,1)$ or a $K(Z_p,1)$.

Another generalization of the lens space may be obtained as follows. Let $k\leq n-1$, and let $q_1,q_2,\cdots,q_k$ be positive integers relatively prime to p. Define an equivalence relation on $\mathbf{S}^{2n-1}$ by $(z_1,\cdots,z_n)\sim(\zeta z_1,\cdots,\zeta z_{n-k},\zeta^{q_1}z_{n-k+1},\cdots,\zeta^{q_k}z_n)$, where ζ is a primitive pth root of unity. The quotient set is denoted by $\mathbf{L}^n(p,q_1,\cdots,q_k)$, and is called a *generalized lens space*. The cell complex on this set will again have one cell in each dimension m for $0\leq m\leq 2n-1$.

Examples of cellular decompositions of the classical Lie groups and their associated Stiefel manifolds which are similar to those described above may be found in the work of Miller [26], Steenrod [36], and Yokota [43].

Example **2.7.** *Simplicial complexes.* A subset A of a real vector space V is independent if $t_0\alpha_0+t_1\alpha_1+\cdots+t_n\alpha_n\neq 0$, where $\alpha_i\in A$, $t_i\in R$, and $\sum_i t_i=1$. If A is an independent set, the *simplex of A* is the set

$$\Delta(A)=\{\sum_i t_i\alpha_i \mid \alpha_i\in A, t_i\geq 0 \quad \text{and} \quad \sum_i t_i=1\}.$$

The elements of A are called *vertices* of the simplex $\Delta(A)$.

For any set A, the collection of functions $\beta_a:A\to R$ defined by $\beta_a(a')=0$ for $a\neq a'$, $\beta_a(a)=1$ form an independent subset of the real vector space R^A. Clearly the collection of β_a is in one-to-one correspondence with the elements of A. Thus for any set there is a simplex whose vertices are in one-to-one correspondence with the set.

A simplex has a canonical cell structure. First, suppose A is an independent finite set, say $A=\{\alpha_0,\alpha_1,\cdots,\alpha_n\}$. Let $\Delta^n\subset R^{n+1}$ be the standard n-simplex (see 0.1) and define $\varphi_A:\Delta^n\to\Delta(A)$ by $\varphi_A(t_0,t_1,\cdots,t_n)=\sum_i t_i\alpha_i$. Then φ_A is a bijective characteristic map for the n-cell of $\Delta(A)$. In the general case, we choose a characteristic map $\varphi_{A'}$ for each finite subset $A'\subset A$, and let $\Phi_A=\{\varphi_{A'} \mid A'\subset A \text{ is finite}\}$. One easily checks that $(\Delta(A),\Phi_A)$ is a regular cell structure on $\Delta(A)$. A *simplicial complex* is a subcell structure of a simplex. We remark that one does not usually identify under the relation of strict equivalence in the simplicial theory.

Let $\gamma\in\Delta(A)$. Then $\gamma=\sum_{\alpha\in A} t_\alpha\alpha$, where $t_\alpha\geq 0$, $t_\alpha=0$ for all but finitely many $\alpha\in A$, and $\sum_{\alpha\in A} t_\alpha=1$. The number t_α is uniquely determined by γ and is called the *barycentric coordinate* of γ corresponding to the vertex α. The *barycentric coordinate functions* are the functions $b_\alpha:\Delta(A)\to R$ defined by $b_\alpha(\gamma)=t_\alpha$. For a simplicial complex, the barycentric coordinate functions are the restrictions of those of an ambient simplex and are independent of the simplex in which the simplicial complex is embedded.

An *abstract simplicial complex* K, or *vertex scheme*, consists of a set A, called the vertices of K, together with a collection $\mathcal{S}$ of finite subsets of A with the following properties:

(i) for each $a\in A$, the set $\{a\}$ belongs to $\mathcal{S}$;
(ii) if A' belongs to $\mathcal{S}$, then so does each subset.

The elements of $\mathcal{S}$ are called the simplexes of K, and if a simplex contains $n \geq 1$ distinct elements, it is called an $n-1$ simplex. Clearly, there is a one-to-one correspondence between the simplicial complexes embedded in $\Delta(A)$ and the abstract simplicial complexes which have A for a set of vertices.

A simplicial map of one simplex $\Delta(A)$ into a second $\Delta(B)$ is a map $f:A \to B$ which is linearly extended over the subset $\Delta(A)$ of the vector space V to the subset $\Delta(B)$ of the vector space V'. If $f:K \to L$ is a map of one simplicial complex into another which is the restriction of a simplicial map on simplexes containing the two, then it is called a *simplicial map.*

A map $f:K \to L$ of abstract simplicial complexes which maps vertices into vertices and sends the simplex $\{a_0, \cdots, a_n\}$ into $\{f(a_0), \cdots, f(a_n)\}$ is also called a simplicial map. Clearly, the simplicial maps between two simplicial complexes correspond one to one with the simplicial maps between corresponding abstract simplicial complexes.

Simplicial maps are examples of what we will later call regular cellular maps (4.1).

The following are simple pathologies in the definition of cell structure.

Example **2.8.** If X is any set, there is a trivial cell structure on X: one 0-cell σ_x^0 for each point x of X, and corresponding characteristic map $\varphi_x:E^0 \to x \in X$.

Example **2.9.** Let D^2 be the closed 2-disc $\{(x,y) \mid x^2+y^2 \leq 1\}$. A useless cell structure for D^2 is one with one 2-cell (corresponding to all of D^2) and one 0-cell x for each point x in $S^1 = \dot{D}^2$. We will eliminate this type of pathology in Chapter II.

Example **2.10.** The following is an example of a nonnormal cell structure. The set X is the subset of R^3 obtained as a union of a

1- and 2-sphere as shown in Fig. 1.1. It has one 0-cell, one 1-cell (whose interior is the 1-sphere minus σ^0), and one 2-cell (whose interior is the 2-sphere minus the circle). Evidently, the obvious characteristic maps are the extensions to closures of the homeo-

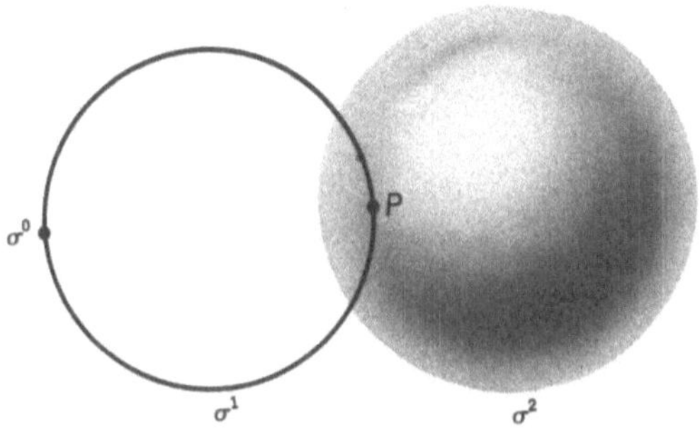

Figure 1.1. The point P is $\dot{\sigma}^2$. The 2-sphere and the circle are tangent at P.

morphisms of open cells onto the interiors of cells. Thus $\dot{E}^1, \dot{E}^2$ go into σ^0 and the point P, respectively. Note that *the closed cells are not subcomplexes*; the smallest subcomplex of X containing the 2-cell is the whole complex.

3. CARRIER THEORY

In this section we define carriers, the carrier topology, and give some of their properties.

Definition **3.1.** The *carrier topology* of a cell complex $(X,\mathcal{S})$ is the topology on X in which the closed sets are the subcomplexes of $(X,\mathcal{S})$. This defines a topology by lemma 1.5. Observe that a "closed cell" need not be closed in this topology, as in example 2.9. A cell is closed in the carrier topology if and only if it is a normal cell.

The closure of a set A is the smallest closed set containing it in the carrier topology, hence is the smallest subcomplex of $(X,\mathcal{S})$ whose underlying set contains A. As usual, this can be described as follows.

Definition **3.2.** If $(X,\mathcal{S})$ is a cell complex, the *carrier* of $A \subset X$ is the intersection of all subcomplexes of $(X,\mathcal{S})$ whose underlying set contains A. The carrier of A will be denoted by $(C(A),\mathcal{S} \mid A)$, or by just $C(A)$ if the cellular decomposition is fixed.

LEMMA **3.3.** *If* $(X,\mathcal{S})$ *is a cell complex,* $A \subset X$, *and* $C(A)$ *is the carrier of* A, *then*

(i) $C(\varnothing) = \varnothing$;

(ii) *for each* $A \subset X$, $A \subset C(A)$;

(iii) *for each* $A \subset X$, $C(C(A)) = C(A)$;

(iv) *for each* $A \subset X$ *and* $B \subset X$, $C(A \cup B) = C(A) \cup C(B)$;

(v) *for each* $A \subset X$ *and* $B \subset X$, *if* $A \subset B$, *then* $C(A) \subset C(B)$.

Proof. Since the carrier topology is a topology, the first four properties follow as in Kelley [22, pp. 42–43]. The fifth property follows from the first four. In fact, these properties are equivalent to the subcomplexes of $(X,\mathcal{S})$ being the closed sets of a topology on X [Kelley, *ibid.*]. ∎

Note that by lemma 1.5 the carrier topology on X has the unusual property that the union of any family of closed sets is closed.

LEMMA **3.4.** *If* $(X,\mathcal{S})$ *is a cell complex and* $A \subset X$, *then*

$$C(A) = \bigcup \{C(\sigma) \mid \sigma \in \mathcal{S} \quad \text{and} \quad (\sigma - \dot{\sigma}) \cap A \neq \varnothing\},$$

the union of all $C(\sigma)$, *where* σ *is any cell with an interior point in* A.

Proof. Since the sets $\sigma - \dot{\sigma}$ partition X, we see

$$A \subset \bigcup \{\sigma \mid \sigma \in \mathcal{S} \quad \text{and} \quad (\sigma - \dot{\sigma}) \cap A \neq \varnothing\}.$$

By lemma 3.3, (iv) and (v),

$$C(A) \subset \mathbf{U}\{C(\sigma) \mid \sigma \in \mathcal{S} \quad \text{and} \quad (\sigma - \dot{\sigma}) \cap A \neq \emptyset\}.$$

On the other hand, if $\sigma \in \mathcal{S}$ is such that $(\sigma - \dot{\sigma}) \cap A \neq \emptyset$, then $(\sigma - \dot{\sigma})$ meets any subcomplex of $(X,\mathcal{S})$ which contains A. By 1.6, σ must be contained in such a complex, so that $\sigma \subset C(A)$. Thus $C(\sigma) \subset C(C(A)) = C(A)$, and we obtain

$$\mathbf{U}\{C(\sigma) \mid \sigma \in \mathcal{S} \quad \text{and} \quad (\sigma - \dot{\sigma}) \cap A \neq \emptyset\} \subset C(A). \quad \blacksquare$$

COROLLARY **3.5.** *If* $x \in \sigma - \dot{\sigma}$, *then* $C(x) = C(\sigma - \dot{\sigma}) = C(\sigma)$. ∎

LEMMA **3.6.** *If* $(X,\mathcal{S})$ *is a cell complex and* $\sigma \in \mathcal{S}$, *then*

$$C(\sigma) = C(\dot{\sigma}) \cup (\sigma - \dot{\sigma}).$$

Proof. Clearly $C(\dot{\sigma}) \subset C(\sigma)$, and since $\sigma - \dot{\sigma} \subset \sigma$, we have $(\sigma - \dot{\sigma}) \cup C(\dot{\sigma}) \subset C(\sigma)$. On the other hand, $\sigma = (\sigma - \dot{\sigma}) \cup \dot{\sigma} \subset (\sigma - \dot{\sigma}) \cup C(\dot{\sigma})$, and $(\sigma - \dot{\sigma}) \cup C(\dot{\sigma})$ is the set of a subcomplex of $(X,\mathcal{S})$. Thus $C(\sigma) \subset (\sigma - \dot{\sigma}) \cup C(\dot{\sigma})$. ∎

LEMMA **3.7.** *The cell complex* $(X,\mathcal{S})$ *is closure finite if and only if the carrier of each cell is a finite subcomplex.*

Proof. Assume $(X,\mathcal{S})$ is closure finite, let σ be a cell, and argue inductively on the dimension n of σ. If $n = 0$, then $C(\sigma^0) = \sigma^0$, which is a finite subcomplex. If the lemma is true for dimensions less than n and σ is an n-cell, then $C(\sigma) = (\sigma - \dot{\sigma}) \cup C(\dot{\sigma}) = \sigma \cup C(\dot{\sigma})$, and we only have to show that $C(\dot{\sigma})$ is a finite complex. But $C(\dot{\sigma}) = \mathbf{U}\{C(\tau) \mid (\tau - \dot{\tau}) \cap \dot{\sigma} \neq \emptyset\}$ is contained in X^{n-1}, so that each τ such that $(\tau - \dot{\tau}) \cap \dot{\sigma} \neq \emptyset$ has dimension $q < n$ and by hypothesis this is a finite union. By the inductive hypothesis, each $C(\tau)$ is a finite complex, and thus $C(\dot{\sigma})$ must be finite.

Conversely, if $(X,\mathcal{S})$ is not closure finite, then for some cell σ^n there are infinitely many cells τ^q with $q < n$ such that $\sigma^n \cap (\tau^q - \dot{\tau}^q) \neq \emptyset$, and therefore by 3.4, $C(\sigma^n)$ is infinite. ∎

Corollary 3.8. *The cell complex* $(X,\mathcal{S})$ *is closure finite if and only if each point lies in a finite subcomplex (has finite carrier).* ∎

Lemma 3.9. *For any cell* σ *of* $(X,\mathcal{S})$, *the following three statements are equivalent:*

(1) *the cell* σ *is normal (i.e.,* σ *carries the structure of a subcomplex, or* $\sigma = C(\sigma)$);

(2) *the boundary* $\dot\sigma$ *carries the structure of a subcomplex (i.e.,* $C(\dot\sigma) = \dot\sigma$);

(3) $C(\dot\sigma) \subset \sigma$.

Proof. (1) implies (2): If σ is closed in the carrier topology, so is $\dot\sigma = \sigma \cap X^{n-1}$, the intersection of two closed sets. (2) implies (3): $\dot\sigma = C(\dot\sigma)$ implies that $C(\dot\sigma) = \dot\sigma \subset \sigma$. (3) implies (1): $C(\dot\sigma) \subset \sigma$ implies that $C(\sigma) = (\sigma - \dot\sigma) \cup C(\dot\sigma) \subset \sigma$ and therefore that $C(\sigma) = \sigma$. ∎

Definition 3.10. A cell complex $(X,\mathcal{S})$ is *connected* if it is connected in the carrier topology.

As usual, each component of $(X,\mathcal{S})$ is closed, i.e., is a subcomplex of $(X,\mathcal{S})$. Because the carrier topology has the property that the union of closed sets is closed, the components of $(X,\mathcal{S})$ are open as well.

Lemma 3.11. *If* $(X,\mathcal{S})$ *is a cell complex and* $\sigma \in \mathcal{S}$, *then* $\sigma - \dot\sigma$ *is connected in the relative carrier topology.*

Proof. Suppose $(\sigma - \dot\sigma) \cap A \cap B = \emptyset$ and $\sigma - \dot\sigma \subset A \cup B$, where A and B are closed in $(X,\mathcal{S})$. Then A and B are the spaces of subcomplexes of $(X,\mathcal{S})$, and, by 1.2, either $\sigma - \dot\sigma \subset A$ or $\sigma - \dot\sigma \subset B$, i.e., either $(\sigma - \dot\sigma) \cap A = \emptyset$ or $(\sigma - \dot\sigma) \cap B = \emptyset$. ∎

Corollary 3.12. *If* $(X,\mathcal{S})$ *is a cell complex and* $\sigma \in \mathcal{S}$, *then* σ *and* $C(\sigma)$ *are both connected.*

Proof. Since $C(\sigma - \dot\sigma) = C(\sigma)$ by 3.5, $\sigma - \dot\sigma \subset \sigma \subset C(\sigma - \dot\sigma)$. But

$C(\sigma - \dot{\sigma})$ is the closure of a connected set, hence is connected. Also any set between a connected set and its closure is connected, so that σ is connected. ∎

Definitions **3.13.** Let $(X,\mathcal{S})$ be a cell complex. A *combinatorial path* or *c-path* of $(X,\mathcal{S})$ is a sequence $(x_0,x_1, \cdots, x_n)$ of points of X such that there exist cells $\sigma_i \in \mathcal{S}$ for which $\{x_{i-1},x_i\} \subset \sigma_i$, $i = 1, 2, \cdots, n$. We say that x_0 is the *initial point*, x_n is the *terminal point*, and x_0, x_n are the *endpoints* of the c-path $(x_0,x_1, \cdots, x_n)$. We also say that the c-path $(x_0,x_1, \cdots, x_n)$ is a c-path from x_0 to x_n. If $p = (x_0,x_1, \cdots, x_n)$ and $q = (x_n,x_{n+1}, \cdots, x_{n+m})$ are c-paths of $(X,\mathcal{S})$ we define $p \vee q = (x_0,x_1, \cdots, x_n,x_{n+1}, \cdots, x_{n+m})$. Clearly this is a c-path of $(X,\mathcal{S})$ which we call the *product* of the c-paths p and q. It is worth noting that a product between the c-paths p and q is defined if and only if p and q have an end-point in common; and when any triple product of c-paths p, q, and r is defined, it is associative. If $p = (x_0,x_1, \cdots, x_n)$ is a c-path of $(X,\mathcal{S})$, we write $p^{-1} = (x_n,x_{n-1}, \cdots, x_0)$ and call p^{-1} the *inverse* of p. Note that $p \vee p^{-1}$ and $p^{-1} \vee p$ are always defined. The c-path $\epsilon_x = (x)$ is called the *constant* c-path at $x \in X$. Note that if $p = (x_0,x_1, \cdots, x_n)$, then $\epsilon_{x_0} \vee p = p$ and $p \vee \epsilon_{x_n} = p$. Finally, the c-path $p = (x_0,x_1, \cdots, x_n)$ is in $A \subset X$ provided each of the cells σ_i is a subset of A.

LEMMA **3.14.** *If $(X,\mathcal{S})$ is a cell complex, $\sigma \in \mathcal{S}$, $x \in \sigma - \dot{\sigma}$, and $y \in C(\sigma)$, there is a c-path in $C(\sigma)$ from x to y.*

Proof. Recall that $C(\sigma) = (\sigma - \dot{\sigma}) \cup C(\dot{\sigma})$. If $y \in \sigma - \dot{\sigma}$, then (x,y) is a c-path in $C(\sigma)$ from x to y, because $\{x,y\} \subset \sigma \subset C(\sigma)$. Now suppose that $y \in C(\dot{\sigma}^n)$. If $n = 1$, $C(\dot{\sigma}^n) = \dot{\sigma}^n \subset \sigma^n = C(\sigma^n)$, and therefore $\{x,y\} \subset \sigma^n \subset C(\sigma^n)$ and (x,y) is a c-path in $C(\sigma^n)$ from x to y. Suppose the theorem is true for all cells $\sigma^k \in \mathcal{S}$ with $k \leq n-1$. By lemma 3.4,

$$C(\dot{\sigma}^n) = \bigcup \{C(\tau) \mid \tau \in \mathcal{S} \quad \text{and} \quad (\tau - \dot{\tau}) \cap \dot{\sigma}^n \neq \varnothing\} \subset X^{n-1};$$

thus if $y \in C(\dot{\sigma}^n)$, $y \in C(\tau^q)$ with $q \leq n-1$ and there exists $z \in (\tau^q - \dot{\tau}^q) \cap \dot{\sigma}^n$. By the induction hypothesis, there is a c-path p in $C(\tau^q)$ from z to y, and since $z \in \sigma^n$, $\{x,z\} \subset \sigma^n \subset C(\sigma^n)$. Thus (x,z)

is a c-path in $C(\sigma^n)$ and the product $(x,z)\vee p$ is a c-path in $C(\sigma^n)$ from x to y. ∎

Proposition 3.15. *Let $(X,\mathcal{S})$ be a cell complex, let $C_x \subset X$ be the component of $x \in X$ in the carrier topology, and let $K_x = \{y \mid$ there is a c-path from x to $y\}$. Then $C_x = K_x$.*

Proof. By lemma 3.4, if $y \in C(K_x)$, then $y \in C(\sigma)$ for some cell σ such that there is a $z \in (\sigma - \dot{\sigma}) \cap K_x$. Since $z \in K_x$, there is a c-path q in $C(\sigma)$ from z to y. The product $p \vee q$ is a c-path from x to y. Thus $C(K_x) \subset K_x$, and since $K_x \subset C(K_x)$, $K_x = C(K_x)$, and we conclude that K_x is closed. Now suppose that $y \in C(X - K_x) \cap K_x$. By lemma 3.4, $y \in C(\sigma)$ for some $\sigma \in \mathcal{S}$ such that there is a $z \in (\sigma - \dot{\sigma}) \cap (X - K_x)$. By lemma 3.14, there is a c-path q from y to $z \in X - K_x$. But $y \in K_x$, so that there is a c-path p from x to y, and the product $p \vee q$ is a c-path from x to z. This is impossible by definition of K_x, hence $C(X - K_x) \subset X - K_x$, and we conclude that $X - K_x$ is also closed. Thus K_x is both open and closed, and we conclude that the component $C_x \subset K_x$, because $x \in C_x \cap K_x$.

Now suppose that $y \in K_x$, $x = x_0$, $y = x_n$, and $(x_0, x_1, \cdots, x_n)$ is a c-path from x to y. Let $\sigma_i \in \mathcal{S}$ be such that $\{x_{i-1}, x_i\} \subset \sigma_i$ for $i = 1, 2, \cdots, n$. Now by corollary 3.12, σ_i is connected for $i = 1, 2, \cdots, n$. Since $x \in \sigma_1 \cap C_x$, and C_x is a component, we have $\sigma_i \subset C_x$, and $x_1 \in C_x$. If $x_1, x_2, \cdots, x_i \in C_x$, then $\sigma_{i+1} \subset C_x$, and $x_{i+1} \in C_x$. By induction $y = x_n \in C_x$, and $K_x \subset C_x$. We conclude that $C_x = K_x$. ∎

Lemma 3.16. *If $(X,\mathcal{S})$ is a cell complex and $\sigma \in \mathcal{S}$, then $C(\sigma) \cap X^0 \neq \varnothing$.*

Proof. If $n = 0$, $C(\sigma^n) = \sigma^n \subset X^0$. Suppose that for $k \leq n-1$, $C(\sigma^k) \cap X^0 \neq \varnothing$. Since $C(\sigma^n) = (\sigma^n - \dot{\sigma}^n) \cup C(\dot{\sigma}^n)$ and $C(\tau^q) \subset C(\dot{\sigma}^n)$ for some cell τ^q with $q \leq n-1$, we see that $\varnothing \neq C(\tau^q) \cap X^0 \subset C(\dot{\sigma}^n) \cap X^0 \subset C(\sigma^n) \cap X^0$. ∎

Proposition 3.17. *If $(X,\mathcal{S})$ is a cell complex and X^k is connected in the carrier topology for some $k \geq 1$, then X is connected in the carrier topology.*

Proof. Let $x,y \in X$, let $u \in C(x) \cap X^0$, and let $v \in C(y) \cap X^0$. Let p be a c-path in $C(x)$ from x to u, let q be a c-path in X^k from u to v, and let r be a c-path in $C(y)$ from v to y. Then $p \vee q \vee r$ is a c-path in X from x to y. ∎

We remark that the 0-skeleton X^0 is totally disconnected in the carrier topology, and a cell complex can be connected in the carrier topology without X^1 being connected. We will see later that if we put a sufficiently fine topology on X, then X connected implies X^1 is connected.

PROPOSITION **3.18.** *A connected locally finite cell complex* $(X,\mathcal{S})$ *is countable.*

Proof. For each nonnegative integer n and a fixed $\sigma_0 \in \mathcal{S}$, let $A_n = \{(\sigma_0,\sigma_1, \cdots, \sigma_n) \mid \sigma_i \in \mathcal{S} \text{ and } \sigma_{i-1} \cap \sigma_i \neq \varnothing\}$, and define a function $\alpha_n : A_n \to \mathcal{S}$ by $\alpha_n(\sigma_0,\sigma_1, \cdots, \sigma_n) = \sigma_n$. Clearly A_0 is finite, and if A_{n-1} is finite, then $\alpha_{n-1}(A_{n-1})$ is finite. Hence, since $(X,\mathcal{S})$ is locally finite, each A_n is a finite set, and $A = \bigcup_n A_n$ is countable. If $\sigma \in \mathcal{S}$, there is a c-path from a point of σ_0 to a point of σ because $(X,\mathcal{S})$ is connected, so that each cell of $\mathcal{S}$ is in $\alpha_n(A_n)$ for some n. Thus the map $\alpha : A \to \mathcal{S}$ defined by the α_n is a surjection and $\mathcal{S}$ is a countable set. ∎

Because of the peculiar property of the carrier topology that any union of closed sets is closed, the intersection of any family of open sets (complements of subcomplexes) is open. Hence, for any subset of a cell complex we can define a smallest open set which contains it in the same way that we define the closure of a subset.

Definition **3.19.** Let $(X,\mathcal{S})$ be a cell complex and let A be a subset of X. The *star* or *open star* of A, written $\mathrm{st}(A)$, is the complement of the union of all cell carriers $C(\sigma)$ which do not meet A. The *closed star* $\overline{\mathrm{st}}(A)$ is the union of all cell carriers which do meet A.

Since the open star is the complement of a subcomplex, it is an open set, and since it contains A, the open star of A is a neighborhood of A in the carrier topology. The open star is the intersection of all open sets containing A, hence is the unique minimal open set containing A. The closed star of A contains the open star and hence is a neighborhood of A, and being a subcomplex is closed.

LEMMA 3.20. *If $A \subset X$, then $C(\mathrm{st}(A)) = \overline{\mathrm{st}}(A)$, i.e., the closed star of A is the closure of* $\mathrm{st}(A)$ *in the carrier topology.*

Proof. Note that $(\sigma - \dot{\sigma}) \cap \mathrm{st}(A) = \varnothing$ if and only if $\sigma - \dot{\sigma} \subset \bigcup \{C(\tau) \mid C(\tau) \cap A = \varnothing\}$. Since the latter is a subcomplex, this is equivalent to $C(\sigma) \cap A = \varnothing$. Thus

$$
\begin{aligned}
C(\mathrm{st}(A)) &= \bigcup \{C(\sigma) \mid (\sigma - \dot{\sigma}) \cap \mathrm{st}(A) \neq \varnothing\} \\
&= \bigcup \{C(\sigma) \mid C(\sigma) \cap A \neq \varnothing\} \\
&= \overline{\mathrm{st}}(A). \quad \blacksquare
\end{aligned}
$$

LEMMA 3.21. *Let $(X, \mathcal{S})$ be a normal cell complex. Then* $\mathrm{st}(A)$ *is the union of all open cells $\sigma - \dot{\sigma}$ such that σ meets A, and* $\overline{\mathrm{st}}(A)$ *is the union of all closed cells that meet A.*

Proof. Since $(X, \mathcal{S})$ is normal, $C(\sigma) = \sigma$ for each cell σ. Thus $\overline{\mathrm{st}}(A) = \bigcup \{\sigma \mid \sigma \cap A \neq \varnothing\}$. Let $S = \bigcup \{\sigma - \dot{\sigma} \mid \sigma \cap A \neq \varnothing\}$, and observe that $A \subset S$. The set $X - S$ is a subcomplex of X, for if $(\tau - \dot{\tau}) \cap (X - S) \neq \varnothing$, then $\tau \cap A = \varnothing$, hence $\tau = C(\tau) \subset X - A \subset X - S$. Thus $X - S$ is closed, and S is open in the carrier topology and contains A, so that $S \supset \mathrm{st}(A)$. On the other hand, if $\sigma \cap A \neq \varnothing$, then σ belongs to no subcomplex $K \subset X$ such that $K \cap A = \varnothing$. Consequently, $\sigma - \dot{\sigma} \subset X - \bigcup \{\tau \mid \tau \cap A = \varnothing\} = \mathrm{st}(A)$. $\blacksquare$

4. FUNCTIONS

As usual, one wants to describe a category which has as objects the cell complexes. We discuss some appropriate types of maps.

Definitions **4.1.** Let $(X,\mathcal{S})$ and $(Y,\mathfrak{J})$ be cell complexes. A function $f:X\to Y$ is said to be *cellular* provided that, for each n, $f(X^n)\subset Y^n$. A function $f:X\to Y$ is *regular* if f is cellular, and for each $\sigma^n\in\mathcal{S}$, $f(\sigma^n)=\tau^m\in\mathfrak{J}$ and $f(\sigma^n-\dot{\sigma}^n)=\tau^m-\dot{\tau}^m$. Of course, if f is regular and $f(\sigma^n)=\tau^m$, then $m\leq n$ because f is cellular.

Example **4.2.** If $(X,\mathcal{S})$ and $(Y,\mathfrak{J})$ are the cell complexes of simplicial complexes, then a simplicial map $f:X\to Y$ is a regular cellular map.

Example **4.3.** Let $(\mathbf{S}^n,\mathcal{S})$ be the regular cell complex of example 2.2, and let $T:\mathbf{S}^n\to\mathbf{S}^n$ be the antipodal map. Then T is a regular cellular map.

Example **4.4.** Let $(\mathbf{S}^n,\mathcal{S})$ be the regular cell complex of example 2.2, and let $(\mathbf{RP}^n,\mathfrak{J})$ be the cell complex of example 2.4. The map $\pi_R:\mathbf{S}^n\to\mathbf{RP}^n$ is a regular cellular map.

PROPOSITION **4.5.** *If $(X,\mathcal{S})$ is a cell complex, the identity map $1_X:X\to X$ is a regular cellular map.*

PROPOSITION **4.6.** *If $(X,\mathcal{S})$, $(Y,\mathfrak{J})$, and $(Z,\mathfrak{U})$ are cell complexes and if $f:X\to Y$ and $g:Y\to Z$ are regular cellular maps, then the composition $gf:X\to Z$ is a regular cellular map.*

PROPOSITION **4.7.** *If $(X,\mathcal{S})$ and $(Y,\mathfrak{J})$ are cell complexes and $f:X\to Y$ is a regular cellular function which is a bijection, then $f^{-1}:Y\to X$ is a regular cellular function.*

The proofs of these three propositions are very easy and are left to the reader. ∎

Definition **4.8.** If $(X,\mathcal{S})$ and $Y,\mathfrak{J})$ are cell complexes and if there is a regular cellular bijection $f:X\to Y$, we say that $(X,\mathcal{S})$ and $(Y,\mathfrak{J})$ are *isomorphic*.

PROPOSITION **4.9.** *If* $(X,\mathfrak{S})$ *and* $(Y,\mathfrak{T})$ *are cell complexes and* $f:X\to Y$ *is a regular cellular map, then* f *is continuous and closed in the carrier topologies on* X *and* Y.

Proof. Let $C_{\mathfrak{S}}$ and $C_{\mathfrak{T}}$ denote the carriers on X and Y, respectively. Suppose that $\sigma\in\mathfrak{S}$, $\tau'\in\mathfrak{T}$, and $(\sigma-\dot{\sigma})\cap f^{-1}(C_{\mathfrak{T}}(\tau'))\neq\emptyset$. Then $f(\sigma-\dot{\sigma})=\tau-\dot{\tau}$ for some $\tau\in\mathfrak{T}$ such that $(\tau-\dot{\tau})\cap C_{\mathfrak{T}}(\tau')\neq\emptyset$. But then $\tau\subset C_{\mathfrak{T}}(\tau')$ because $C_{\mathfrak{T}}(\tau')$ is a subcomplex, and therefore we see that $\sigma\subset f^{-1}(C_{\mathfrak{T}}(\tau'))$. This means that $f^{-1}(C_{\mathfrak{T}}(\tau'))$ is a subcomplex of X, so that $C_{\mathfrak{S}}(f^{-1}(\tau'))\subset f^{-1}(C_{\mathfrak{T}}(\tau'))$, and it follows from lemma 3.3 that for any $B\subset Y$, $C_{\mathfrak{S}}(f^{-1}(B))\subset f^{-1}(C_{\mathfrak{T}}(B))$, so that f is continuous.

To see that f is closed, let $A\subset X$ be a subcomplex, and suppose that $(\tau-\dot{\tau})\cap f(A)\neq\emptyset$ for some $\tau\in\mathfrak{T}$. Let $x\in(\tau-\dot{\tau})\cap f(A)$. Then $x=f(y)$ for some $y\in A$, and $y\in\sigma-\dot{\sigma}\subset A$ for some $\sigma\in\mathfrak{S}$. Now $f(\sigma-\dot{\sigma})$ is an open cell of Y which meets $\tau-\dot{\tau}$, and therefore $\tau=f(\sigma)\subset f(A)$. Thus $f(A)$ is a subcomplex of Y and f is closed. ∎

COROLLARY **4.10.** *If* $(X,\mathfrak{S})$ *and* $(Y,\mathfrak{T})$ *are isomorphic, then they are homeomorphic in the carrier topologies.*

Proof. If $f:X\to Y$ is a regular cellular bijection, then f^{-1} is also, and since f is closed and continuous, f^{-1} is continuous. ∎

COROLLARY **4.11.** *If* $(X,\mathfrak{S})$ *and* $(Y,\mathfrak{T})$ *are cell complexes and* $f:X\to Y$ *is a regular cellular map,* f *maps components of* X *to components of* Y. ∎

Definition **4.12.** Let X, Y be cell complexes and $\mathfrak{C}$ a collection of subsets of X. A *$\mathfrak{C}$-carrier* E from X to Y is a function from $\mathfrak{C}$ to the set of subcomplexes of Y such that if $A\subset B$, and both are in $\mathfrak{C}$, then $E(A)\subset E(B)$. If E and E' are $\mathfrak{C}$-carriers from X to Y and for each $A\in\mathfrak{C}$ we have $E(A)\subset E'(A)$, we say E is *dominated* by E' and write $E\subset E'$. If E is a $\mathfrak{C}$-carrier from X to Y and $f:X\to Y$ is a map, we say f *is carried by* E if for each $A\in\mathfrak{C}$, we have $f(A)\subset E(A)$.

LEMMA. 4.13 *Let $f:X \to Y$ be a map of cell complexes, and let $\mathcal{C}$ be any collection of subsets of X. Define $E(A) = C(f(A))$ for $A \in \mathcal{C}$. Then E is a $\mathcal{C}$-carrier, and for any other $\mathcal{C}$-carrier E' carrying f, we have $E' \supset E$. We call E the minimal $\mathcal{C}$-carrier of f.*

The proof is left as an exercise. ∎

For example, let $(X, \mathcal{S})$ be a cell complex and $\mathcal{C}$ be all the subsets of X, and let $1_X : X \to X$ be the identity map. Then the minimal $\mathcal{C}$-carrier of 1_X is the carrier function sending A into $C(A)$.

5. PRODUCT COMPLEXES

Let $(X, \mathcal{S})$ and $(Y, \mathcal{J})$ be cell complexes and Φ, Φ' representative collections of characteristic maps. Let σ^n be an n-cell of $\mathcal{S}$, τ^m an m-cell of $\mathcal{J}$, and φ, φ' their characteristic maps in Φ, Φ'. Finally, let $\varphi \times \varphi' : E^n \times E^m \to X \times Y$ be the product map onto $\sigma^n \times \tau^m$. Note that $\varphi \times \varphi'(\dot{E}^{n+m}) = \varphi \times \varphi'((E^n \times \dot{E}^m) \cup (\dot{E}^n \times E^m)) = (\sigma^n \times \dot{\tau}^m) \cup (\dot{\sigma}^n \times \tau^m)$. Let $\Phi \times \Phi'$ be the collection of all maps $\varphi \times \varphi'$ for $\varphi \in \Phi$ and $\varphi' \in \Phi'$.

PROPOSITION 5.1. *The pair $(X \times Y, \Phi \times \Phi')$ is a cell structure. The equivalence class of $\Phi \times \Phi'$ depends only on $\mathcal{S}$ and $\mathcal{J}$, and we may therefore denote it by $\mathcal{S} \times \mathcal{J}$.*

Proof. Since $E^{n+m} - \dot{E}^{n+m} = E^n \times E^m - (E^n \times E^m)^{\bullet} = (E^n - \dot{E}^n) \times (E^m - \dot{E}^m)$, and the product of bijective maps is bijective, each map $\varphi \times \varphi'$ is bijective on the interior of its domain cell. Since the product of partitions on X and Y is a partition of $X \times Y$, the sets $\varphi \times \varphi'(E^{n+m} - \dot{E}^{n+m}) = (\sigma^n - \dot{\sigma}^n) \times (\tau^m - \dot{\tau}^m)$ partition $X \times Y$. The map $\varphi \times \varphi'$ sends $\dot{E}^{n+m}$ into $(\sigma^n \times \dot{\tau}^m) \cup (\dot{\sigma}^n \times \tau^m) \subset (X^n \times Y^{m-1}) \cup (X^{n-1} \times Y^m)$, and therefore $\varphi \times \varphi'(\dot{E}^{n+m})$ is contained in a union of images of the form $\psi \times \psi'(E^q)$, where $q < n+m$ and $\psi \in \Phi$, $\psi' \in \Phi'$. Thus $\Phi \times \Phi'$ determines a cell structure on $X \times Y$.

Clearly, if we change each map φ and φ' by a reparametrization, a homeomorphism of the corresponding domain cells E^n and E^m, then we only change the product map by a reparametrization of E^{n+m}. ∎

Definition **5.2.** If $(X,\mathcal{S})$ and $(Y,\mathfrak{I})$ are cell complexes, the cell complex $(X\times Y,\ \mathcal{S}\times\mathfrak{I})$ is called the *product cell complex.* We point out that the n-skeleton of the product is the set

$$(X\times Y)^n = \bigcup_p (X^p \times Y^{n-p}).$$

By induction one may define a product cell complex for any finite set of cell complexes, but there is no natural way to define infinite products. One easily sees that the properties of finiteness, countability, local finiteness, and local countability all carry over to finite products of cell complexes with these properties. Since the product of bijections is a bijection, a finite product of regular cell complexes is a regular cell complex.

PROPOSITION **5.3.** *Let* $(X,\mathcal{S})$ *and* $(Y,\mathfrak{I})$ *be cell complexes and let* $\pi_X : X\times Y \to X$ *and* $\pi_Y : X\times Y\to Y$ *be the projection maps. Then* π_X *and* π_Y *are regular cellular maps from the cell complex* $(X\times Y,\ \mathcal{S}\times\mathfrak{I})$ *to* $(X,\mathcal{S})$ *and* $(Y,\mathfrak{I})$*, respectively.*

The proof is easy and will be left as an exercise. ∎

PROPOSITION **5.4.** *If* $(A,\mathcal{S}')$ *and* $(B,\mathfrak{I}')$ *are subcomplexes of the cell complexes* $(X,\mathcal{S})$ *and* $(Y,\mathfrak{I})$*, respectively, then* $(A\times B,\ \mathcal{S}'\times\mathfrak{I}')$ *is a subcomplex of* $(X\times Y,\ \mathcal{S}\times\mathfrak{I})$.

Proof. Suppose $\sigma\times\tau$ is a cell of $\mathcal{S}\times\mathfrak{I}$ whose interior meets $A\times B$. Then

$$\varnothing \neq ((\sigma\times\tau)-(\sigma\times\tau)^{\cdot})\cap(A\times B) = ((\sigma-\dot{\sigma})\times(\tau-\dot{\tau})\cap(A\times B))$$

$$= ((\sigma-\dot{\sigma})\cap A)\times((\tau-\dot{\tau})\cap B).$$

Thus we must have $\sigma\subset A$ and $\tau\subset B$, and therefore $\sigma\times\tau\subset A\times B$. ∎

COROLLARY **5.5.** *If* $(X,\mathcal{S})$ *and* $(Y,\mathfrak{I})$ *are closure finite or normal cell complexes, then so is* $(X\times Y,\ \mathcal{S}\times\mathfrak{I})$.

Proof. If the cells σ and τ lie in finite subcomplexes, or carry the structure of subcomplexes, then the same is true for $\sigma\times\tau$. ∎

PROPOSITION **5.6.** *If* $(X,\mathcal{S})$ *and* $(Y,\mathfrak{I})$ *are cell complexes and* $A\subset X$, $B\subset Y$, *then the carrier satisfies* $C(A\times B)=C(A)\times C(B)$.

Proof. We first prove this in the case where $A\times B\subset(X\times Y)^n$. The proof is by induction on n. If $A\times B\subset X^0\times Y^0=(X\times Y)^0$, then A, B and $A\times B$ consist of 0-cells only, so that $C(A\times B)=A\times B=C(A)\times C(B)$. Suppose the proposition is true for all $A\times B\subset(X\times Y)^k$ with $0\leq k\leq n-1$. Let $p+q=n$ and let σ be a p-cell of $\mathcal{S}$ and τ a q-cell of $\mathfrak{I}$. Then we have

$$C(\sigma\times\tau)=((\sigma\times\tau)-(\sigma\times\tau)^{\cdot})\cup C((\sigma\times\tau)^{\cdot})$$

$$=((\sigma-\dot{\sigma})\times(\tau-\dot{\tau}))\cup C(\sigma\times\dot{\tau})\cup C(\dot{\sigma}\times\tau)$$

$$=((\sigma-\dot{\sigma})\times(\tau-\dot{\tau}))\cup(C(\sigma)\times C(\dot{\tau}))\cup(C(\dot{\sigma})\times C(\tau))$$

by the induction hypothesis, since $\sigma\times\dot{\tau}$ and $\dot{\sigma}\times\tau$ are subsets of $(X\times Y)^{n-1}$. If we now write $C(\sigma)=(\sigma-\dot{\sigma})\cup C(\dot{\sigma})$ and $C(\tau)=(\tau-\dot{\tau})\cup C(\dot{\tau})$, we see that $C(\sigma\times\tau)=C(\sigma)\times C(\tau)$. Thus

$$C(A\times B)=\bigcup\{C(\sigma\times\tau)\mid((\sigma\times\tau)-(\sigma\times\tau)^{\cdot})\cap(A\times B)\neq\varnothing\}$$

$$=\bigcup\{C(\sigma)\times C(\tau)\mid(\sigma-\dot{\sigma})\cap A\neq\varnothing\text{ and }(\tau-\dot{\tau})\cap B\neq\varnothing\}$$

$$=C(A)\times C(B).$$

In particular, this proves that for a cell $\sigma\times\tau\in\mathcal{S}\times\mathfrak{I}$, $C(\sigma\times\tau)=C(\sigma)\times C(\tau)$.

Now if A and B are any subsets of X and Y, respectively,

$$C(A\times B)=\bigcup\{C(\sigma\times\tau)\mid((\sigma\times\tau)-(\sigma\times\tau)^{\cdot})\cap(A\times B)\neq\varnothing\}$$

$$=\bigcup\{C(\sigma)\times C(\tau)\mid(\sigma-\dot{\sigma})\cap A\neq\varnothing\text{ and }(\tau-\dot{\tau})\cap B\neq\varnothing\}$$

$$=C(A)\times C(B).$$ ∎

Corollary **5.7.** *If* $(X,\mathcal{S})$ *and* $(Y,\mathfrak{I})$ *are cell complexes, the carrier topology on* $(X\times Y,\ \mathcal{S}\times\mathfrak{I})$ *is the product topology for the carrier topologies on* $(X,\mathcal{S})$ *and* $(Y,\mathfrak{I})$. ▮

Corollary **5.8.** *If* $(X,\mathcal{S})$ *and* $(Y,\mathfrak{I})$ *are cell complexes, the components of* $(X\times Y,\ \mathcal{S}\times\mathfrak{I})$ *are the products of the components of* $(X,\mathcal{S})$ *and* $(Y,\mathfrak{I})$. ▮

6. EQUIVALENCE RELATIONS AND QUOTIENTS

Many of the most useful constructions of topology are obtained by identifying various subsets of some space, for example, the cones, suspensions, mapping cylinders, etc. A familiar construction of this type on cell complexes is the one that yields the compact 2-dimensional manifolds as identification spaces of polygonal cells. This section describes a framework for these constructions in the combinatorial theory.

If X is a set and $\mathcal{R}$ is an equivalence relation on X, let $p:X\to X/\mathcal{R}$ be the quotient map onto the set of equivalence classes $X/\mathcal{R}$.

Definition **6.1.** Let $(X,\mathcal{S})$ be a cell complex and $\mathcal{R}$ an equivalence relation on X. Then $\mathcal{R}$ is a *cellular equivalence relation* provided the following are satisfied.

(i) If $\sigma\in\mathcal{S}$, then $p^{-1}p(\sigma-\dot\sigma)$ is a union of open cells $\sigma'-\dot\sigma'$ of the cellular partition of X.

(ii) If $\sigma'-\dot\sigma'\subset p^{-1}p(\sigma-\dot\sigma)$ is of minimal dimension among all such open cells in the union, then $p\mid(\sigma'-\dot\sigma')$ is a bijection onto $p(\sigma-\dot\sigma)$ and $p(\sigma')=p(\sigma)$. Such a cell σ' will be called *$\mathcal{R}$-minimal* for the cell σ.

(iii) If σ' and σ'' are both $\mathcal{R}$-minimal for the cell σ and if φ' and φ'' are the respective characteristic functions, then there is a homeomorphism $h:E_{\sigma'}\to E_{\sigma''}$ such that $p\varphi'=p\varphi''h$.

Theorem **6.2.** *Let* $(X,\mathcal{S})$ *be a cell complex and* $\mathcal{R}$ *a cellular equivalence relation on* X. *Define* $\mathcal{S}/\mathcal{R}=\{p(\sigma)\mid\sigma\in\mathcal{S}$ *and* σ *is* $\mathcal{R}$*-minimal*$\}$. *Then* $(X/\mathcal{R},\ \mathcal{S}/\mathcal{R})$ *is a cell complex.*

Proof. For each set $p(\sigma) \in \mathcal{S}/\mathcal{R}$ choose one minimal cell, say $\sigma \in \mathcal{S}$, and choose a characteristic map $\varphi_\sigma : E_\sigma \to \sigma$. With these choices, let $\Phi/\mathcal{R}$ be the set of maps $p\varphi_\sigma$. We first show that $(X/\mathcal{R}, \Phi/\mathcal{R})$ is a cell structure.

(i) Since σ was chosen to be a minimal cell, it is clear that $p\varphi_\sigma \mid (E_\sigma - \dot{E}_\sigma)$ is a bijection.

(ii) Each point of $X/\mathcal{R}$ is of the form $p(x)$ for some $x \in X$. Suppose $x \in \tau - \dot{\tau}$, and σ is a minimal cell for τ. Then $p(x) \in p(\sigma - \dot{\sigma})$ and $p(\sigma) \in \mathcal{S}/\mathcal{R}$. Thus the sets $p(\sigma - \dot{\sigma})$ cover $X/\mathcal{R}$. Suppose that $p(\sigma - \dot{\sigma}) \cap p(\tau - \dot{\tau}) \neq \varnothing$ for minimal cells σ, τ. Then $p^{-1}p(\sigma - \dot{\sigma}) \cap p^{-1}p(\tau - \dot{\tau}) \neq \varnothing$, and since each of these are unions of open cells which partition X, they have a cell in common. Hence $p^{-1}p(\sigma - \dot{\sigma}) = p^{-1}p(\tau - \dot{\tau})$, and $p(\sigma - \dot{\sigma}) = p(\tau - \dot{\tau})$. But we chose exactly one minimal cell for $p(\sigma) \in \mathcal{S}/\mathcal{R}$, so $\sigma = \tau$. Thus the sets $p(\sigma - \dot{\sigma})$ for $p(\sigma) \in \mathcal{S}/\mathcal{R}$ partition $X/\mathcal{R}$.

(iii) For the minimal cell σ, we know that $\varphi_\sigma(\dot{E}_\sigma) = \dot{\sigma}$, and the cells contained in $\dot{\sigma}$ all have smaller dimension than σ. Thus their minimal cells will have dimension less than that of σ, so that $p\varphi_\sigma(\dot{E}_\sigma)$ is contained in a union of cells from $\mathcal{S}/\mathcal{R}$ of dimension less than the dimension of σ.

This completes the proof that $(X/\mathcal{R}, \Phi/\mathcal{R})$ is a cell structure, and we only need check that our choices work correctly with respect to strict equivalence. But any other choice of minimal cells and characteristic maps gives a strictly equivalent cell structure by property (iii) of the definition of cellular equivalence relation. ∎

Definition **6.3.** If $(X,\mathcal{S})$ is a cell complex and $\mathcal{R}$ is a cellular equivalence relation on X, the complex $(X/\mathcal{R}, \mathcal{S}/\mathcal{R})$ is the *quotient* or *identification* complex of $(X,\mathcal{S})$ with respect to $\mathcal{R}$.

Proposition **6.4.** *The quotient map $p : X \to X/\mathcal{R}$ is a regular cellular map.*

Proof. Since $p(\sigma) = p(\sigma')$, where σ' is $\mathcal{R}$-minimal for σ, and since σ' and $p(\sigma')$ have the same dimension which is no greater than the dimension of σ, p maps the n-skeleton of X into the n-skeleton of $X/\mathcal{R}$. From the definition of the cellular structure on $X/\mathcal{R}$ and condition (ii), p is regular. ∎

COROLLARY **6.5.** *If* $(X,\mathcal{S})$ *is closure finite or normal, then the same is true of a quotient complex with respect to a cellular equivalence.*

Proof. The set X has a closure finite cell structure if and only if each cell lies in a finite subcomplex. Since $p:X\to X/\mathcal{R}$ is a regular cellular map, the image of a subcomplex of X is a subcomplex of $X/\mathcal{R}$. Therefore, since every cell of $X/\mathcal{R}$ is the image of a cell of X, every cell of $X/\mathcal{R}$ lies in a finite subcomplex.

The set $X/\mathcal{R}$ has a normal cell structure if and only if each cell is a subcomplex. If this is true for X, it will be true for $X/\mathcal{R}$ because p is a regular cellular map. ∎

COROLLARY **6.6.** *If* $\mathcal{R}$ *is a cellular equivalence relation, the quotient map* $p:X\to X/\mathcal{R}$ *is continuous and closed in the carrier topologies. Thus if* $(X,\mathcal{S})$ *is connected, so is* $(X/\mathcal{R}, \mathcal{S}/\mathcal{R})$.

Proof. Simply apply proposition 4.9 and the standard theorem that the continuous image of a connected set is connected. ∎

If $\mathcal{R}$ and $\mathcal{R}'$ are equivalence relations on the sets X and X', respectively, a map $f:X\to X'$ is *$\mathcal{R}$-$\mathcal{R}'$-compatible* if $(x,y)\in\mathcal{R}$ implies $(f(x),f(y))\in\mathcal{R}'$. If the map $f:X\to X'$ is $\mathcal{R}$-$\mathcal{R}'$-compatible, then f induces a map $f:X/\mathcal{R}\to X'/\mathcal{R}'$.

PROPOSITION **6.7.** *If* $\mathcal{R}$ *and* $\mathcal{R}'$ *are cellular equivalence relations on the cell complexes* $(X,\mathcal{S})$ *and* $(X',\mathcal{S}')$, *respectively, and if* $f:X\to X'$ *is a (regular) cellular map which is* $\mathcal{R}$-$\mathcal{R}'$*-compatible, then the induced map* $\bar{f}:X/\mathcal{R}\to X'/\mathcal{R}'$ *is a (regular) cellular map.*

Proof. Let $p:X\to X/\mathcal{R}$ and $p':X'\to X'/\mathcal{R}'$ be the quotient maps. If $x\in (X/\mathcal{R})^n$, then $x=p(y)\in p(\sigma-\dot{\sigma})$ for some $\mathcal{R}$-minimal cell σ of dimension at most n, i.e., $y\in X^n$. Since $\bar{f}(x)=\bar{f}p(y)=p'f(y)$

and since f is cellular, $f(y) \in X'^n$. Thus $f(y) \in \tau - \dot{\tau}$ for some cell τ of dimension at most n, and $\bar{f}(x) \in p'(\tau - \dot{\tau})$. Therefore, $\bar{f}(x) \in (X'/\mathcal{R}')^n$, and $\bar{f}$ is cellular.

If f is regular as well, let $\bar{\sigma}$ be a cell of $X/\mathcal{R}$, and let $\sigma - \dot{\sigma} \subset p^{-1}(\bar{\sigma} - \dot{\bar{\sigma}})$. Since f and p' are regular cellular maps, if $p'f(\sigma) = \tau$ and $p'f(\sigma - \dot{\sigma}) = \tau - \dot{\tau}$, then $f(\bar{\sigma}) = \tau$ and $f(\bar{\sigma} - \dot{\bar{\sigma}}) = \tau - \dot{\tau}$. ∎

In the remainder of this section we give some examples of cellular equivalence relations and some of the constructions based on them.

PROPOSITION **6.8.** *Let* $(X,\mathcal{S})$ *be a cell complex and* $(A_\gamma,\mathfrak{I}_\gamma)$ *a family of disjoint subcomplexes. If* Δ *is the diagonal of* $X \times X$ *and* $\mathcal{R} = \Delta \cup \bigcup_\gamma (A_\gamma \times A_\gamma)$, *then* $\mathcal{R}$ *is a cellular equivalence relation.* ∎

We say the quotient complex $(X/\mathcal{R}, \mathcal{S}/\mathcal{R})$ obtained from this relation is obtained from $(X,\mathcal{S})$ by *shrinking* or *collapsing the subcomplexes* $(A_\gamma,\mathfrak{I}_\gamma)$ *to vertices of* $(X, \mathcal{S})$.

In the following examples $I = (I,\mathcal{I})$ will denote the standard regular cell complex on the unit interval I consisting of two 0-cells (denoted by 0 and 1 and called the endpoints of I) and one 1-cell which has the identity map as a characteristic map. If $X = (X,\mathcal{S})$ is a cell complex, the following are all cell complexes.

(1) The *cone over* X, $c(X)$, is obtained from $X \times I$ by collapsing the subcomplex $X \times \{1\}$ to a vertex.

(2) The *reduced cone over* X, $\bar{c}(X)$, is obtained from $X \times I$ by collapsing the subcomplex $(X \times \{1\}) \cup (\{\bar{x}\} \times I)$ to a vertex, where $\bar{x}$ is a vertex (0-dimensional subcomplex) of X. One may also obtain it from the cone by collapsing a generator, the image of the subcomplex $\{\bar{x}\} \times I$ in $c(X)$.

(3) The *suspension of* x, $s(X)$, is obtained from $X \times I$ by collapsing the subcomplexes $X \times \{0\}$ and $X \times \{1\}$ to distinct vertices.

(4) *The reduced suspension of* X, $\bar{s}(X)$, is obtained from $X \times I$ by collapsing the subcomplex $(X \times \{0\}) \cup (\{\bar{x}\} \times I) \cup (X \times \{1\})$ to a point, where $\bar{x}$ is a vertex of X.

(5) If $(X,\mathcal{S})=X$ and $(Y,\mathfrak{I})=Y$ are cell complexes, the *reduced product* or *smash product* $X\wedge Y$ is obtained from $X\times Y$ by collapsing the subcomplex $(X\times\{\bar{y}\})\cup(\{\bar{x}\}\times Y)$ to a point. Here $\bar{x}$ and $\bar{y}$ are vertices of X and Y, respectively.

In each of these examples, if both X and Y are normal or closure finite, then so is the resulting complex, since products and quotients preserve these properties.

Let $(X,\mathcal{S})$ be a cell complex and $\{(A_\gamma,\mathfrak{I}_\gamma)\mid\gamma\in\Gamma\}$ a family of subcomplexes. Let Ω be a family of regular cellular bijections $\omega_{\gamma',\gamma}:A_\gamma\to A_{\gamma'}$ (not necessarily defined for all pairs $(\gamma,\gamma')\in\Gamma\times\Gamma$) which satisfies:

(a) if $\omega\in\Omega$, then $\omega^{-1}\in\Omega$;
(b) for each $\gamma\in\Gamma$, $\omega_{\gamma,\gamma}\in\Omega$ and is the identity map of A_γ;
(c) if $\omega_{\gamma',\gamma}\in\Omega$ and $\omega_{\gamma'',\gamma'}\in\Omega$, then the composition $\omega_{\gamma'',\gamma'}\omega_{\gamma',\gamma}=\omega_{\gamma'',\gamma}\in\Omega$;
(d) $\omega_{\gamma,\gamma'}\mid A_\gamma\cap A_{\gamma'}$ is either the identity, or $(\sigma-\dot{\sigma})\cap\omega_{\gamma,\gamma'}(\sigma-\dot{\sigma})=\varnothing$;
(e) if $\omega\in\Omega$, $\omega(\sigma)=\sigma'$, and φ_σ, $\varphi_{\sigma'}$ are characteristic maps, there is a homeomorphism $h:E_\sigma\to E_{\sigma'}$ such that $\omega\varphi_\sigma=\varphi_{\sigma'}h$.

Then Ω is called a *family of identifications* on X.

Proposition 6.9. *If $(X,\mathcal{S})$ is a cell complex, Ω is a family of identifications on X, and if $\mathcal{R}_\Omega=\Delta\cup\{(x,x')\in X\times X\mid$ there is an $\omega\in\Omega$ such that $\omega(x)=x'\}$, then $\mathcal{R}_\Omega$ is a cellular equivalence relation.*

Proof. Clearly $\mathcal{R}_\Omega$ is an equivalence relation. Let $p:X\to X/\mathcal{R}_\Omega$ be the quotient map.

(i) If $(\sigma-\dot{\sigma})\cap A_\gamma=\varnothing$ for all γ, then $p^{-1}p(\sigma-\dot{\sigma})=\sigma-\dot{\sigma}$. If $(\sigma-\dot{\sigma})\cap A_\gamma\neq\varnothing$ for some γ and $z\in p^{-1}p(\sigma-\dot{\sigma})$, $z=\omega(x)$ for some $\omega\in\Omega$ and some $x\in\sigma-\dot{\sigma}$. Suppose $z\in\tau-\dot{\tau}$. Since ω is a regular cellular map, $\omega(\sigma-\dot{\sigma})=\tau-\dot{\tau}$. Thus $\tau-\dot{\tau}\subset p^{-1}p(\sigma-\dot{\sigma})$. We have proved that $p^{-1}p(\sigma-\dot{\sigma})$ is a union of open cells.
(ii) Since the maps $\omega\in\Omega$ are regular cellular and $\omega^{-1}\in\Omega$ whenever

$\omega \in \Omega$, if σ is an n-cell, then $\omega(\sigma) = \tau$ is an n-cell. Thus $p^{-1}p(\sigma - \dot{\sigma})$ is a union of open n-cells, and each open cell of $p^{-1}p(\sigma - \dot{\sigma})$ is $\mathcal{R}_\Omega$-minimal. Since $\omega_{\gamma,\gamma'} \mid A_\gamma \cap A_{\gamma'}$ is either the identity or $(\sigma - \dot{\sigma}) \cap \omega_{\gamma,\gamma'}(\sigma - \dot{\sigma}) = \varnothing$, $p \mid (\sigma - \dot{\sigma})$ is a bijection onto $p(\sigma - \dot{\sigma})$.

(iii) This is clearly satisfied by condition (d). ∎

The following are some examples of complexes which may be obtained by using a family of identifications.

(1) Let $(I,\mathcal{J})$ be the standard complex on the unit interval, and let Ω be a family of identifications of the subcomplexes $\{0\}$ and $\{1\}$ which identify $\{0\}$ and $\{1\}$. The resulting identification complex is the cell complex on the 1-sphere $\mathbf{S}^1$ given in 2.1.

(2) Let P be a $2n$-sided regular polygon in the plane, say with vertices $\{v_1, v_2, \cdots, v_{2n}\}$ and edges $\{e_1, e_2, \cdots, e_{2n}\}$. Let D be the compact region bounded by P, and give D a cellular structure so that there is one 2-cell σ, $2n$ 1-cells e_i, and $2n$ 0-cells v_i, with the obvious characteristic maps. For $i = 1, 2, \cdots, n$, let $\omega_i : e_i \to e_{i+n}$ be a homeomorphism, and let Ω consist of the ω_i, their inverses, and the identity maps for all the cells e_i. Then Ω is a family of identifications and $D/\mathcal{R}_\Omega$ is one of the compact 2-manifolds. In fact, any compact 2-manifold except the 2-sphere $\mathbf{S}^2$ can be obtained in this manner. See Massey [23, pp. 1–34] for details.

(3) The construction of the lens spaces $L^n(p,q)$ as given in 2.6.

(4) Another cellular decomposition of the suspension of a space can be given by taking as X the disjoint union of two cones $c^+(X) \cup c^-(X)$, and as Ω the cellular bijections which identify the two bases of the cones, together with the appropriate identity maps. A similar construction works for the reduced suspension.

7. *ADJUNCTION COMPLEXES*

Let X and Y be sets, let $A \subset X$, and let $f : A \to Y$ be a function. Define the set $Y \cup_f X$ as the quotient of the disjoint union of X and Y under the equivalence relation which identifies $a \in A$ with

$f(a) \in Y$. The set $Y \cup_f X$ is called the *adjunction of X to Y by the map f*. Let $p: Y \cup X \to Y \cup_f X$ be the quotient map. We point out that the subsets Y and $X - A$ of the disjoint union $Y \cup X$ are mapped bijectively into $Y \cup_f X$.

THEOREM 7.1. *Let (X, Φ) and (Y, Ψ) be cell structures, let (A, Φ_A) be a substructure of (X, Φ), and let $f: A \to Y$ be a cellular map. Then if $\Psi \cup_f \Phi = \{p\psi \mid \psi \in \Psi\} \cup \{p\varphi \mid \varphi \in \Phi - \Phi_A\}$, the pair $(Y \cup_f X, \Psi \cup_f \Phi)$ is a cell structure. Moreover, the strict equivalence class of this cell structure is determined by the class of the cell structures (X, Φ) and (Y, Ψ) and the function f.*

Proof. (i) If $\theta \in \Psi \cup_f \Phi$, then either $\theta = p\varphi$, where $\varphi(E^n - \dot{E}^n) \subset X - A$, or $\theta = p\psi$. In either case θ is bijective on the interior of the domain cell.

(ii) Clearly $X - A$ is partitioned by the sets $\varphi(E^n - \dot{E}^n)$ where $\varphi \in \Phi - \Phi_A$ and Y is partitioned by the sets $\psi(E^n - \dot{E}^n)$ for $\psi \in \Psi$. Since $Y \cup_f X$ is partitioned by the sets $p(X - A)$ and $p(Y)$, we see that $Y \cup_f X$ is partitioned by the sets $\theta(E^n - \dot{E}^n)$.

(iii) If $\theta = p\varphi$, let $B_1 = \varphi^{-1}(\varphi(\dot{E}^n) \cap (X - A))$ and $B_2 = \varphi^{-1}(\varphi(\dot{E}^n) \cap A)$. Then $\varphi(B_1) \subset X^{n-1} \cap (X - A)$, and $\theta(B_1) \subset p(X^{n-1} \cap (X - A))$. Also $\theta(B_2) = p(\varphi(\dot{E}^n) \cap A) = f(\varphi(\dot{E}^n) \cap A) \subset Y^{n-1}$ because f is cellular. Thus $\theta(\dot{E}^n)$ is contained in a union of cells of dimension less than n. If $\theta = p\psi$, then $\psi(\dot{E}^n) \subset Y^{n-1}$ because (Y, Ψ) is a cell structure.

This completes the first part of the theorem. The statement about strict equivalence is very easy and will be omitted. ∎

Definition 7.2. We denote the cell complex obtained from cell complexes $(X, \mathfrak{S})$ and $(Y, \mathfrak{J})$ by adjunction of X to Y as above by $(Y \cup_f X, \mathfrak{S} \cup_f \mathfrak{J})$ and call it the *adjunction complex*. It is worth pointing out that, although the adjunction set $Y \cup_f X$ is obtained from an equivalence relation, the adjunction complex is not usually a quotient complex in the sense of section 6, even if we restrict f to be a regular cellular map.

The following properties are easily verified and their proofs will be omitted.

PROPOSITION **7.3.** *Let* $(Y\cup_f X,\ \mathcal{S}\cup_f \mathfrak{I})$ *be an adjunction complex for a cellular map* $f:A\to Y$ *and let* $p:Y\cup X\to Y\cup_f X$ *be the quotient map. Then*

(i) $\mathfrak{I}\cup_f \mathcal{S}=\{p(\tau)\mid \tau\in\mathfrak{I}\}\cup\{p(\sigma)\mid \sigma\in\mathcal{S}$ *and* $\sigma-\dot{\sigma}\subset X-A\}$;

(ii) *the complex* $(Y,\mathfrak{I})$ *is isomorphic to the subcomplex* $(p(Y),\{p(\tau)\mid \tau\in\mathfrak{I}\})$ *of the adjunction complex under the map* $p\mid Y$;

(iii) *if* $(B,\mathcal{S}_B)$ *is any subcomplex of* $(X,\mathcal{S})$ *such that* $B\subset X-A$, *then* $(B,\mathcal{S}_B)$ *is isomorphic to a subcomplex of the adjunction under the map* $p\mid B$. ▮

Under the conditions of (ii) and (iii) of this proposition, we usually identify $(Y,\mathfrak{I})$ and $(B,\mathcal{S}_B)$ with their images in the adjunction complex and say that $(Y,\mathcal{S})$ and $(B,\mathcal{S}_B)$ are subcomplexes of the adjunction.

PROPOSITION **7.4.** *If* $(X,\mathcal{S})$ *and* $(Y,\mathfrak{I})$ *are closure finite cell complexes,* $(A,\mathcal{S}_A)$ *is a subcomplex of* $(X,\mathcal{S})$ *and* $f:A\to Y$ *is a cellular map such that* $C(f(\sigma))$ *is a finite complex for each* $\sigma\in\mathcal{S}$, *then the adjunction complex* $(Y\cup_f X,\ \mathfrak{I}\cup_f \mathcal{S})$ *is a closure finite complex.*

Proof. Suppose that $\sigma-\dot{\sigma}\subset X-A$, and σ meets the lower dimensional cells σ_i for $i=1, 2, \cdots, n$. Assume that $\sigma_i-\dot{\sigma}_i\subset X-A$ for $i=1, 2, \cdots, k$ and $\sigma_i\subset A$ for $i=k+1, k+2, \cdots, n$. Then $p(\sigma)\subset \sigma_1\cup\sigma_2\cup\cdots\cup\sigma_k\cup C(f(\sigma_{k+1}))\cup\cdots\cup C(f(\sigma_n))$ and this union contains finitely many cells of dimension less than the dimension of σ. Thus $p(\sigma)$ meets finitely many cells of lower dimension. If $\tau\in\mathfrak{I}$, $p(\tau)$ meets only the cells $p(\tau')$, where τ meets τ' in Y. Since $(Y,\mathfrak{I})$ is closure finite, $p(\tau)$ meets only finitely many cells of lower dimension. ▮

In the following $(I,\mathcal{g})$ is the standard cell complex on the unit interval I, and $(X,\mathcal{S})$ and $(Y,\mathfrak{I})$ are cell complexes.

(1) The *mapping cylinder*. Let $f:X\to Y$ be a cellular map. We adjoin $X\times I$ to Y by the map $\bar{f}:X\times\{1\}\to Y$ defined by $\bar{f}(x,1)=f(x)$. The resulting complex $(M_f,\mathfrak{M}_f)$ is called the mapping

cylinder. It contains $(Y,\mathfrak{I})$ as a subcomplex, and since $(X,\mathcal{S})$ is isomorphic to $(X\times\{0\}, \mathcal{S}\times\{0\})$ which is a subcomplex of $(M_f,\mathfrak{M}_f)$, we say $(X,\mathcal{S})$ is a subcomplex of $(M_f,\mathfrak{M}_f)$.

(2) The *join.* Define $f\colon (X\times\{0\}\times Y)\cup(X\times\{1\}\times Y)\to X\cup Y$ (disjoint union) by $f(x,0,y)=x$ and $f(x,1,y)=y$. Then the complex obtained by adjoining $X\times I\times Y$ to $X\cup Y$ by f is a complex called the join of X and Y and is usually denoted by $(X*Y, \mathcal{S}*\mathfrak{I})$.

(3) The *cone* and *reduced cone.* Let $\{1\}$ be a one point complex, and define $f\colon X\times\{1\}\to\{1\}$. The complex obtained by adjoining $X\times I$ to $\{1\}$ by f is the cone $c(X)$. To get the reduced cone $\bar{c}(X)$, one adjoins $X\times I$ to $\{1\}$ by the function $f\colon (X\times\{1\})\cup(\{\bar{x}\}\times I)\to\{1\}$, where $\bar{x}$ is a vertex of X.

(4) The *suspension* and *reduced suspension.* Let $\dot{I}=\{0,1\}$ be a two point complex and define $f\colon X\times\dot{I}\to\dot{I}$ by $f(x,0)=0$ and $f(x,1)=1$. The suspension $s(X)$ is obtained by adjoining $X\times I$ to $\dot{I}$ by f. To get the reduced suspension $\bar{s}(X)$, adjoin $X\times I$ to the one point complex $\{1\}$ by the map $f((X\times\{0\})\cup(\{\bar{x}\}\times I)\cup(X\times\{1\}))=1$, where $\bar{x}$ is a vertex of X.

(5) The *smash product.* Adjoin $X\times Y$ to the one point complex $\{1\}$ by the map $f((X\times\{\bar{y}\})\cup\{\bar{x}\}\times Y))=1$, where $\bar{x}$ and $\bar{y}$ are vertices of X and Y, respectively.

CHAPTER II

CW COMPLEXES

Our problem now is to relate the cellular structure of the cell complex $(X,\mathcal{S})$ to a topology on the set X. The result is called a CW complex. The concept of a CW complex and many of the results of this chapter about such a space are due to J. H. C. Whitehead [41].

Choose a set of characteristic functions $\{\varphi_\sigma \mid \sigma \in \mathcal{S}\}$ for the cells of $\mathcal{S}$. Then the *weak topology on X with respect to $\mathcal{S}$* is obtained by:

(a) giving each cell $\sigma \in \mathcal{S}$ the quotient topology with respect to its characteristic function;

(b) giving X the weak topology with respect to the subsets $\sigma \in \mathcal{S}$, i.e., a set $F \subset X$ is closed if and only if $F \cap \sigma$ is closed in σ for each $\sigma \in \mathcal{S}$.

Recall that a cell complex $(X,\mathcal{S})$ is closure finite if each cell $\sigma \in \mathcal{S}$ meets finitely many cells of lower dimension (see I.1.7 and I.3.7).

1. DEFINITIONS

Definition **1.1.** A Hausdorff space X is a *CW complex* with respect to a family of cells $\mathcal{S}$ provided:

(i) the pair $(X,\mathcal{S})$ is a cell complex such that each cell $\sigma \in \mathcal{S}$ has a continuous characteristic function;

(ii) the space X has the weak topology with respect to $\mathcal{S}$;

(iii) the cell complex $(X,\mathcal{S})$ is closure finite.

If $(X,\mathcal{S})$ is a cell complex such that X satisfies (i), (ii), and (iii) with respect to $\mathcal{S}$, we usually say that X is a *CW complex with cells* $\mathcal{S}$. Except in this first section, we will not mention the family of cells unless it is explicitly needed.

Property (i) relates the topology of each cell $\sigma \in \mathcal{S}$ to the topology of the more familiar space the Euclidean cell E_σ. In fact, we have the lemma:

Lemma 1.2. *Let X be a CW complex with cells $\mathcal{S}$. Then*
(1) each cell $\sigma \in \mathcal{S}$ is a closed subset of X;
(2) for each cell $\sigma \in \mathcal{S}$, the restriction of the characteristic map φ_σ to $E_\sigma - \dot{E}_\sigma$ is a homeomorphism onto $\sigma - \dot{\sigma}$.

Proof. The characteristic map φ_σ for $\sigma \in \mathcal{S}$ is a closed map; for if $K \subset E_\sigma$ is any closed set, K is compact because E_σ is. Since φ_σ is continuous, $\varphi_\sigma(K)$ is compact in X, and since X is Hausdorff, $\varphi_\sigma(K)$ is closed in X. Taking $K = E_\sigma$, we get part (1). Since φ_σ is closed, its restriction $\varphi_\sigma \mid (E_\sigma - \dot{E}_\sigma)$ is a closed continuous bijection onto $\sigma - \dot{\sigma}$, hence a homeomorphism. ∎

Property (ii) relates the topology of X to that of the cells $\sigma \in \mathcal{S}$ and yields the following important proposition.

Proposition 1.3. *If X is a CW complex with cells $\mathcal{S}$, if Y is a space, and if $f: X \to Y$ is a function, then f is continuous if and only if the restriction $f \mid \sigma$ is continuous for each $\sigma \in \mathcal{S}$.*

Proof. Suppose that $f: X \to Y$ is continuous and $C \subset Y$ is closed. Then $f^{-1}(C)$ is closed in X, and $(f \mid \sigma)^{-1}(C) = f^{-1}(C) \cap \sigma$ is closed in σ for each cell σ. Thus $f \mid \sigma$ is continuous for each σ.

Suppose that $f: X \to Y$ is such that $f \mid \sigma$ is continuous for each cell σ and $C \subset Y$ is a closed set. Then $(f \mid \sigma)^{-1}(C)$ is closed in σ for each σ, and therefore $f^{-1}(C)$ is closed in X. ∎

Property (iii) is more subtle, and as we shall see immediately, together with (i) and (ii), it forces the following desirable properties of CW complexes.

Proposition 1.4. *Let X be a CW complex with cells $\mathcal{S}$, and let $(A,\mathcal{J})$ be a subcomplex of $(X,\mathcal{S})$. Then*
(1) the set A is closed in X;
(2) in the relative topology A is a CW complex with cells $\mathcal{J}$.

Proof. (1) Let $\sigma \in \mathcal{S}$ and consider the set $A \cap \sigma$. If A contains an interior point of σ, then $\sigma \subset A$ by I.1.6, and $A \cap \sigma = \sigma$ is closed in σ. Also, if $A \cap \sigma = \emptyset$, then $A \cap \sigma$ is closed in σ, and we are interested in the case $\emptyset \neq A \cap \sigma = A \cap \dot{\sigma}$. Since $(X,\mathcal{S})$ is closure finite, σ meets finitely many open cells of lower dimension. Let $\tau_1, \tau_2, \cdots, \tau_k$ be cells of dimension less than that of σ such that $(\tau_i - \dot{\tau}_i) \cap (\sigma \cap A) \neq \emptyset$. Then $\dot{\sigma} \cap A \subset \dot{\sigma} \cap (\mathbf{U}_i\, \tau_i)$ and $\mathbf{U}_i\, \tau_i \subset A$, so that $\sigma \cap (\mathbf{U}_i\, \tau_i) \subset \sigma \cap A = \dot{\sigma} \cap A \subset \dot{\sigma} \cap (\mathbf{U}_i\, \tau_i) = \sigma \cap (\mathbf{U}_i\, \tau_i)$. By 1.2 (1) each τ_i is closed in X, and the finite union $\mathbf{U}_i\, \tau_i$ is closed in X. Thus $\sigma \cap A = \sigma \cap (\mathbf{U}_i\, \tau_i)$ is closed in σ. Since X has the weak topology with respect to $\mathcal{S}$, A is closed in X.

Since the subspace $A \subset X$ is Hausdorff and properties (i) and (iii) are obviously inherited by the subcomplex $(A,\mathcal{J})$, to prove (2) we need only check property (ii) for $(A,\mathcal{J})$. If A is given the relative topology and F is closed in A, then F is closed in X because, by (1), A is closed. But then $F \cap \sigma$ is closed in σ for each $\sigma \in \mathcal{S}$, and, in particular, $F \cap \sigma$ is closed in σ for each $\sigma \in \mathcal{J}$. Suppose that $F \subset A$ is such that $F \cap \tau$ is closed in τ for each $\tau \in \mathcal{J}$. Since $F \subset A$, for each $\sigma \in \mathcal{S} - \mathcal{J}$, $\sigma \cap F = \dot{\sigma} \cap F$. Let $\tau_1, \tau_2, \cdots, \tau_k \in \mathcal{J}$ be the cells of dimension less than the dimension of σ such that $(\tau_i - \dot{\tau}_i) \cap \dot{\sigma} \cap F \neq \emptyset$. Then each set $\tau_i \cap F$ is closed in τ_i, hence in X, so that $\mathbf{U}_i\, (\tau_i \cap F)$ is closed in X. But then $\sigma \cap F = \dot{\sigma} \cap F = \dot{\sigma} \cap (\mathbf{U}_i\, (\tau_i \cap F))$ is closed in σ. Thus F is closed in X. ∎

The following theorem gives a property for *CW* complexes which is equivalent to the closure finiteness property (iii). It is useful in applications because it characterizes the compact subsets of a *CW* complex.

Theorem 1.5. *Let X be a Hausdorff space, let $\mathcal{S}$ be a family of cells on X satisfying properties* (i) *and* (ii), *and let C be the carrier function for $(X,\mathcal{S})$. Then $(X,\mathcal{S})$ is closure finite if and only if*

(iii)′ *if $K \subset X$ is compact, then $C(K)$ is a finite subcomplex of X.*

Proof. Suppose $(X,\mathcal{S})$ is closure finite and $K\subset X$ is compact. Since $C(K)=\bigcup\{C(\sigma)\mid(\sigma-\dot{\sigma})\cap K\neq\emptyset\}$, and since by lemma I.3.7, closure finiteness implies the complex $C(\sigma)$ is finite for each $\sigma\in\mathcal{S}$, we only need to prove that the set

$$\mathcal{A}=\{\sigma\in\mathcal{S}\mid(\sigma-\dot{\sigma})\cap K\neq\emptyset\}$$

is finite. For each $\sigma\in\mathcal{A}$ choose one point $x_\sigma\in(\sigma-\dot{\sigma})\cap K$, and let A be the set of such x_σ. Since $\mathcal{A}$ and A are in one to one correspondence, it suffices to prove A is finite. If $B\subset A$ and $\sigma'\in\mathcal{S}$, $B\cap\sigma'\subset B\cap C(\sigma')$ is finite because $C(\sigma')$ contains finitely many cells. Thus B is closed in σ', and since X has the weak topology with respect to $\mathcal{S}$, B is closed in X. But this means A is a discrete subset of the compact set K, hence it must be finite.

Suppose compact sets have finite carrier and $\sigma\in\mathcal{S}$. Since σ is the continuous image of a closed Euclidean cell which is compact, σ is compact. But then $C(\sigma)$ is a finite complex, so that $(X,\mathcal{S})$ is closure finite by lemma I.3.7. ∎

Corollary 1.6. *If X is a CW complex with cells $\mathcal{S}$, then X is compact if and only if $\mathcal{S}$ is finite.*

Proof. If X is compact, then $X=C(X)$ and $\mathcal{S}$ must be finite by the theorem. If $\mathcal{S}=\{\sigma_1,\cdots,\sigma_n\}$, then $X=\sigma_1\cup\sigma_2\cup\cdots\cup\sigma_n$ where each σ_i is compact, and thus X is compact. ∎

2. ALTERNATIVE DESCRIPTIONS OF CW COMPLEXES

We recall the definition of the adjunction space. Let X and Y be spaces, let A be a closed subset of X, and let $f:A\to Y$ be a continuous map. The *adjunction space* $Y\cup_f X$ is the quotient of the disjoint union of Y and X by the smallest equivalence relation which identifies each $a\in A$ with its image $f(a)\in Y$. If we let $\Phi:Y\cup X\to Y\cup_f X$ be the quotient map, then it is easily checked that $\Phi\mid Y$ is a homeomorphism of Y onto the closed subset $\Phi(Y)$, that $\Phi\mid(X-A)$ is a homeomorphism of $X-A$ onto the open subset $\Phi(X-A)$, and that $Y\cup_f X=\Phi(Y)\cup\Phi(X-A)$ is a partition of $Y \cup_f X$. If A is the empty set, the adjunction space is just the disjoint union.

We will be particularly interested in the following situation. Suppose that X is a space, and for each $\lambda \in \Lambda$, there is a map $f_\lambda : \dot{E}_\lambda \to X$ of the boundary of the Euclidean cell E_λ into X. Let $\mathcal{E} = \bigcup \{E_\lambda \mid \lambda \in \Lambda\}$ and $\dot{\mathcal{E}} = \bigcup \{\dot{E}_\lambda \mid \lambda \in \Lambda\}$ be the disjoint unions, and let $\dot{F} = \bigcup_\lambda f_\lambda : \dot{\mathcal{E}} \to X$ be the union map. The adjunction space $X \cup_{\dot{F}} \mathcal{E}$ is said to be obtained by *attaching the cells E_λ to X*. The maps $f_\lambda : \dot{E}_\lambda \to X$ are called the *attaching maps* for the cells E_λ of $X \cup_{\dot{F}} \mathcal{E}$. The adjunction of a family of 0-cells is just the disjoint union.

Proposition **2.1.** *Let X be a Hausdorff space, and suppose $Y = X \cup_{\dot{F}} \mathcal{E}$ is obtained by attaching the cells $\{E_\lambda \mid \lambda \in \Lambda\}$ to X. Then Y is a Hausdorff space.*

Proof. Let $\Phi : X \cup \mathcal{E} \to X \cup_{\dot{F}} \mathcal{E}$ be the quotient map. If $y_1, y_2 \in \Phi(\mathcal{E} - \dot{\mathcal{E}})$ are distinct, there exist unique $z_1, z_2 \in \mathcal{E} - \dot{\mathcal{E}}$ such that $\Phi(z_i) = y_i$. Since $\mathcal{E} - \dot{\mathcal{E}}$ is Hausdorff, we may choose disjoint open sets U_1, U_2 in $\mathcal{E} - \dot{\mathcal{E}}$ such that $y_i \in U_i$. But then $\Phi(U_1)$ and $\Phi(U_2)$ are disjoint open sets containing y_1 and y_2, respectively. If $y_1 = \Phi(z_1)$ for $z_1 \in \mathcal{E} - \dot{\mathcal{E}}$ and $y_2 \in \Phi(X)$, choose disjoint open sets U_1 and U_2 in the normal space $\mathcal{E}$ such that $z_1 \in U_1$ and $\dot{\mathcal{E}} \subset U_2$. Then $\Phi(U_1)$ and $X \cup_{\dot{F}} U_2$ are disjoint open sets in Y containing y_1 and y_2, respectively. Finally, suppose that $y_1, y_2 \in \Phi(X)$ are distinct. Choose disjoint open sets V_i in X such that $x_i \in V_i$ and $\Phi(x_i) = y_i$. Since $\dot{F}$ is continuous, $\dot{F}^{-1}(V_i)$ is open in $\dot{\mathcal{E}}$, and the sets $W_{i,\lambda} = F^{-1}(V_i) \cap \dot{E}_\lambda$ are disjoint and open in $\dot{E}_\lambda$ for each $\lambda \in \Lambda$ and $i = 1, 2$. Assume each cell E_λ is a copy of the Euclidean disc (vectors of length no greater than 1), define $c_\lambda(V_i) = \{tz \mid z \in F^{-1}(V_i) \cap \dot{E}_\lambda \text{ and } 0 < t \leq 1\}$, and $C_i = \bigcup_\lambda c_\lambda(V_i)$ for $i = 1, 2$. One easily checks that the sets C_i are open and disjoint in $\mathcal{E}$ and that $C_i \cap \dot{\mathcal{E}} = F^{-1}(V_i)$. Now define $U_i = \Phi(C_i - \dot{\mathcal{E}}) \cup V_i$ for $i = 1, 2$. Inasmuch as the sets U_i are disjoint and $y_i \in U_i$, we only need prove that they are open. Since $\Phi^{-1}(U_i) \cap X = V_i$ is open in X and $\Phi^{-1}(U_i) \cap \mathcal{E} = (C_i - \dot{\mathcal{E}}) \cup F^{-1}(V_i) = C_i$ is open in $\mathcal{E}$, the sets U_i are open. ∎

Proposition **2.2.** *Let X be a CW complex, and for each $\lambda \in \Lambda$, let $f_\lambda : \dot{E}_\lambda{}^n \to X^{n-1}$ be a map attaching the cell $E_\lambda{}^n$ to the $(n-1)$-skeleton of X. Define $\mathcal{E}^n = \bigcup \{E_\lambda{}^n \mid \lambda \in \Lambda\}$, $\dot{\mathcal{E}}^n = \bigcup\{\dot{E}_\lambda{}^n \mid \lambda \in \Lambda\}$, and let $\dot{F} = \bigcup_\lambda f_\lambda : \dot{\mathcal{E}}^n \to X^{n-1}$ be the union map. Then $X \cup_{\dot{F}} \mathcal{E}^n$ is a CW complex.*

Proof. By 2.1 above, X is a Hausdorff space. If $\mathcal{S}$ is the set of cells for X, let $\mathcal{S}' = \{\Phi(\sigma) \mid \sigma \in \mathcal{S}\} \cup \{\Phi(E_\lambda{}^n) \mid \lambda \in \Lambda\}$. The proof that $(X \cup_{\dot{F}} \mathcal{E}^n, \mathcal{S}')$ is a cell complex is very easy and is omitted. The characteristic map for one of the attached cells is the restriction $\Phi \mid E_\lambda{}^n$, and that for one of the cells $\Phi(\sigma)$, $\sigma \in \mathcal{S}$, is the composition $\Phi\varphi_\sigma$, both of which are clearly continuous. If $K \subset X \cup_{\dot{F}} \mathcal{E}^n$ is closed, then clearly $K \cap \sigma'$ is closed in σ' for each $\sigma' \in \mathcal{S}'$. Suppose that $K \cap \sigma'$ is closed in σ' for each $\sigma' \in \mathcal{S}'$. If $\sigma' = \Phi(\sigma)$ for $\sigma \in \mathcal{S}$, $\Phi^{-1}(K) \cap \Phi^{-1}(\sigma')$ is closed in $\Phi^{-1}(\sigma')$, and $\Phi^{-1}(K) \cap \Phi^{-1}(\sigma') \cap X = \Phi^{-1}(K) \cap \sigma$ is closed in σ. Since X has the weak topology with respect to $\mathcal{S}$, $\Phi^{-1}(K) \cap X$ is closed in X. If $\sigma' = \sigma_\lambda = \Phi(E_\lambda{}^n)$, $\Phi^{-1}(K) \cap \Phi^{-1}(\sigma_\lambda)$ is closed in $\Phi^{-1}(\sigma_\lambda)$, $\Phi^{-1}(K) \cap \Phi^{-1}(\sigma_\lambda) \cap E_\mu{}^n$, and in particular $\Phi^{-1}(K) \cap \Phi^{-1}(\sigma_\lambda) \cap E_\lambda{}^n = \Phi^{-1}(K) \cap E_\lambda{}^n$ is closed in $E_\lambda{}^n$, and $\Phi^{-1}(K) \cap \mathcal{E}^n$ is closed in $\mathcal{E}^n$. But then $\Phi^{-1}(K)$ is closed in the disjoint union $X \cup \mathcal{E}^n$, so that K is closed in the adjunction space $X \cup_{\dot{F}} \mathcal{E}^n$. Thus $X \cup_{\dot{F}} \mathcal{E}^n$ has the weak topology with respect to $\mathcal{S}'$. If $\sigma' = \Phi(\sigma)$ for $\sigma \in \mathcal{S}$, then σ meets the interiors of cells in $\Phi(X)$ only, and since X is closure finite, such a cell meets only finitely many interiors of cells of lower dimension. Since $\dot{\sigma}_\lambda = \Phi(\dot{E}_\lambda{}^n)$ is a compact subset of X^{n-1}, its carrier is a finite complex. Thus the carrier of σ_λ is a finite complex, and $X \cup_{\dot{F}} \mathcal{E}^n$ is closure finite. ∎

Lemma **2.3.** *Let X and Y be CW complexes and $f : X \to Y$ a continuous map which is a regular cellular bijection. Then f is a homeomorphism.*

Proof. Since each cell σ of X is compact and Y is Hausdorff, $f \mid \sigma$ is a homeomorphism onto its image cell $\tau \subset Y$. Let $K \subset X$ be a closed set. Then $K \cap \sigma$ is closed in σ and $f \mid \sigma$ is a homeomorphism, hence $f(K \cap \sigma)$ is closed in $f(\sigma) = \tau$. Since f is bijective, $f(K \cap \sigma) = f(K) \cap \tau$, so that $f(K) \cap \tau$ is closed in τ, and $f(K)$ is closed in Y. Thus f is a closed continuous bijection. ∎

The following characterization of CW complexes is very useful in many applications because we may check properties skeleton by skeleton rather than cell by cell.

THEOREM 2.4. *A space X is a CW complex if and only if there is a sequence of closed subspaces $X_0 \subset X_1 \subset \cdots \subset X$ such that $X = \bigcup_n X_n$ and:*

(i) *the set X_0 is discrete;*
(ii) *for each n, X_n is obtained from X_{n-1} by attaching n-cells;*
(iii) *the space X has the weak topology with respect to the closed sets X_n.*

Proof. Suppose that X is a CW complex and $X_n = X^n$, the n-skeleton. Then each X^n is closed in X and $X = \bigcup_n X^n$. To prove (i), suppose $B \subset X^0$. Then $B \cap \sigma$ is finite for each cell σ because the carrier of σ is a finite complex, hence has finitely many vertices. Thus $B \cap \sigma$ is closed in σ for each cell σ, and B is closed in X. For (ii), choose a characteristic map $f_\sigma: E_\sigma \to X$ for each n-cell σ of X, let $\mathcal{E}^n = \bigcup \{E_\sigma \mid \sigma \text{ is an } n\text{-cell of } X\}$ and $\dot{\mathcal{E}}^n = \bigcup \{\dot{E}_\sigma \mid \sigma \text{ is an } n\text{-cell of } X\}$ be the disjoint union spaces, let $F = \bigcup_\sigma f_\sigma: \mathcal{E}^n \to X^n$ be the union map, and let $\dot{F} = F \mid \dot{\mathcal{E}}^n: \dot{\mathcal{E}}^n \to X^{n-1}$. Define a map $G: X^{n-1} \cup \mathcal{E} \to X^n$ by $G = i \cup F$, where $i: X^{n-1} \to X^n$ is the inclusion. Clearly G is continuous and compatible with the identifications of $X^{n-1} \cup_{\dot{F}} \mathcal{E}^n$, hence induces a map $G': X^{n-1} \cup_{\dot{F}} \mathcal{E}^n \to X^n$. By proposition 2.2, $X^{n-1} \cup_{\dot{F}} \mathcal{E}^n$ is a CW complex, and one easily sees that G' is a regular cellular bijection. By lemma 2.3, G' is a homeomorphism. Finally, if K is closed in X, K is closed in each X^n. Suppose that $K \subset X^n$ is closed in X^n for each n. Then $K \cap \sigma$ is closed in σ for each $\sigma \subset X$, and K is closed in X, which proves (iii).

Now suppose that X is a space satisfying (i), (ii), and (iii). Since X_0 is discrete, it is certainly a 0-dimensional CW complex, and by proposition 2.2, we may prove each X_n is an n-dimensional complex by induction on n. Since $X = \bigcup_n X_n$, X is a union of images of Euclidean cells which satisfy the axioms for a cell complex. The characteristic map for an n-cell σ of X is the composition of the attaching map for σ with the inclusion of X_n in X. Since X_n is a CW complex and X has the weak topology with respect to the X_n, these characteristic maps are continuous. If $K \subset X$ is such that $K \cap \sigma$ is closed in σ for each σ, then K is closed in each set X_n, because X_n is a CW complex, thus K is closed in X. If K is closed in X, clearly $K \cap \sigma$ is closed in σ for each cell σ. Finally, X is closure finite because each cell is contained in some X_n which is closure finite. ∎

The following is a generalization of the notion of CW complex. The development of this theory parallels that of CW complexes and will be omitted except for the basic definitions.

Definition **2.5.** If X is a Hausdorff space and A is a closed subspace of X, the pair (X,A) is a *relative CW complex* provided there is a sequence of closed subspaces $A = \bar{A}^{-1} \subset \bar{A}^0 \subset \bar{A}^1 \subset \cdots \subset X$ such that $X = \bigcup_n \bar{A}^n$ and:

(i) for each n, $\bar{A}^n$ is obtained from $\bar{A}^{n-1}$ by attaching n-cells;
(ii) the space X has the weak topology with respect to the subspaces $\bar{A}^n$.

The subspace $\bar{A}^n$ is called the *n-skeleton of X relative to A*. A pair (Y,B) is a *subcomplex* of the relative CW complex (X,A) if (Y,B) is a relative CW complex, $Y \subset X$, and $\bar{B}^n = Y \cap \bar{A}^n$ for each n.

The proof of the following proposition is easy and is left to the reader.

Proposition **2.6.** (i) *If* (X,A) *is a relative CW complex and* $A = \varnothing$, *then* X *is a CW complex.*

(ii) *If X is a CW complex and A is a subcomplex, then* (X,A) *is a relative CW complex.* ∎

3. REMARKS ON THE GENERAL TOPOLOGY OF CW COMPLEXES

Although each closed cell of a CW complex is a closed subset, and each subcomplex is closed, it is not true in general that an arbitrary union of closed cells is closed. For example, let X be the following *CW* complex: $X^1 = I = [0,1]$, closed unit interval with the usual cell structure—two vertices and a 1-cell. At each point $x = 1/n$, adjoin a 2-cell by a characteristic map $f_n: \mathbf{S}^1 \to \{1/n\}$. The union of all the 2-cells is not a closed subset. It follows also that the subset $[0,1]$ of I, the union of all the open cells of X whose closures meet the 0-cell $\{0\}$, is not an open set in X, and thus that any neighborhood of $\{0\}$ must meet cells which do not contain $\{0\}$. Thus, in general, we cannot generalize concepts from the theory of simplicial complexes (like "open star") to the theory of CW complexes simply by replacing "simplex" by "cell." However, the following cases are exceptional.

PROPOSITION **3.1.** *If X is either a locally finite or a normal CW complex, then any union of closed cells is a closed subset of X.*

Proof. If X is normal, each cell is a subcomplex, and therefore each union of cells is also. Hence any such union is closed. If X is locally finite and if C is any union of closed cells, then, for any closed cell σ, the intersection $\sigma \cap C$ is a finite union of intersections of σ with other cells and is therefore closed. Thus C meets each σ in a closed subset and is closed. ▮

One way to generalize definitions from the case of simplicial complexes is to replace "simplex" by "cell carrier" or by "finite subcomplex" in definitions, as was done in the definition of open and closed stars in I.3.12. Then, if A is a subset of a CW complex, its open star is open since the carrier topology is refined by the CW topology and thus $\mathrm{st}(A)$ and $\overline{\mathrm{st}}(A)$ are neighborhoods of A. Of course, if X is a normal CW complex, then any such definition reduces to the corresponding one using cells.

PROPOSITION **3.2.** *The collection $\mathfrak{C}$ of cell carriers of a CW complex X has the following properties:*

(1) the elements of $\mathfrak{C}$ are closed and compact;

(2) the covering $\mathfrak{C}$ dominates X;

(3) any intersection of an element $C \in \mathfrak{C}$ with any union of elements of $\mathfrak{C}$ is equal to the intersection of C with a finite subunion.

The proofs of (1) and (3) are very easy and are left to the reader. Part (2) is a special case of part (1) of the next proposition. ∎

In the sequel, we will want to construct open sets and neighborhoods with specified properties. The following proposition (and its corollaries) will be useful for this purpose.

PROPOSITION **3.3.** *Let X be a CW complex and let $\{K_\gamma \mid \gamma \in \Gamma\}$ be a family of subcomplexes such that $\bigcup_\gamma K_\gamma = X$. Then*

(1) *a subset of X is closed if and only if it meets each K_γ in a closed subset;*

(2) *if A is any subset of X and $A \subset V \subset X$, then V is a neighborhood of A if and only if for each K_γ such that $A \cap K_\gamma \neq \varnothing$, $V \cap K_\gamma$ is a neighborhood of $A \cap K_\gamma$.*

Proof. The proof of part (1) is analogous to the proof of part (iii) of theorem 2.4 and will be omitted.

For (2) we may assume $X = \bigcup\{K_\gamma \mid K_\gamma \cap A \neq \varnothing\}$ because this set contains $X - \bigcup\{K_\gamma \mid K_\gamma \cap A = \varnothing\} = K'$ which is an open neighborhood of A. Thus V is a neighborhood of A if and only if $V \cap K'$ is a neighborhood of A in K'. Suppose that for any K_γ there is a subset U_γ which is open in K and such that $A \cap K_\gamma \subset U_\gamma \subset V \cap K_\gamma$. For each subset $\Delta \subset \Gamma$, let $K_\Delta = \bigcup\{K_\gamma \mid \gamma \in \Delta\}$, and consider the set of all pairs (K_Δ, U_Δ), where U_Δ is open in K_Δ and $\varnothing = A \cap K_\Delta \subset U_\Delta \subset V \cap K_\Delta$. Partially order these pairs by $(K_\Delta, U_\Delta) < (K_{\Delta'}, U_{\Delta'})$ if $K_\Delta \subset K_{\Delta'}$ and $U_\Delta = U_{\Delta'} \cap K_\Delta$. Any linearly ordered family of such pairs has as upper bound the pair consisting of the union of the subcomplexes and the union of the corresponding open sets. By Zorn's lemma, there is a maximal such pair (K^*, U^*). We will show that U^* is open in X.

If U^* is not open, by part (1), there must be a subcomplex K such that $U^* \cap K_\gamma$ is not open in K_γ. Hence K_γ is not a subcomplex of K^*. Since $U^* \cap K \cap K^*$ is relatively open in $K_\gamma \cap K^* \subset K^*$, there is an open set $U'' \subset K_\gamma$ such that $U'' \cap K_\gamma \cap K^* = U^* \cap K_\gamma \cap K^*$, and since $K^* \cap U^* \cap K_\gamma \supset A \cap K^* \cap K_\gamma$, we can choose U'' so that it contains $A \cap K_\gamma$ (namely for any choice of U'' take $U'' \cup W$, where W is any open neighborhood of $A - U''$ in

$K_\gamma - U''$). Finally, since the interior of $V \cap K_\gamma$ in K_γ contains $A \cap K_\gamma$ and $U^* \cap K^* \cap K_\gamma$, we can intersect U'' with this interior to get an open set U_γ in K_γ which satisfies $A \cap K_\gamma \subset U_\gamma \subset V \cap K_\gamma$, and $U_\gamma \cap K^* \cap K_\gamma = U^* \cap K^* \cap K_\gamma$. Then $(K^* \cup K_\gamma, U^* \cup U_\gamma)$ is a larger pair than (K^*, U^*), which is a contradiction. ∎

Corollary **3.4.** *If A is a subset of a CW complex X, a set V is a neighborhood of A if and only if $V \cap X^n$ is a neighborhood of $A \cap X^n$ for each n.* ∎

Corollary **3.5.** *If $\mathcal{C}$ is the collection of all finite subcomplexes or of all cell carriers of a CW complex X, and if $A \subset V \subset X$, then the set V is open or closed in X if and only if it meets each member of $\mathcal{C}$ in an open or closed subset, respectively. The set V is a neighborhood of A if and only if it meets each member of $\mathcal{C}$ in a neighborhood of A.* ∎

The following gives some of the local structure of a CW complex.

Proposition **3.6.** *For any CW complex X, the following are equivalent;*

(1) X is locally finite (respectively locally countable);

(2) each point $x \in X$ has a neighborhood meeting only a finite (respectively countable) number of cells;

(3) X is locally compact (respectively locally σ-compact).

Proof. Suppose that X is locally finite and $x \in \sigma - \dot{\sigma} \subset X$. Since there are only finitely many cells which meet σ, $x \in C(x) \subset C(\sigma) \subset C(\tau)$ for finitely many cells τ. But then $\overline{\text{st}}(x) = \bigcup \{C(\tau) \mid x \in C(\tau)\}$ is a finite complex and a neighborhood of x, and its interior meets at most the finitely many cells $\tau_1, \tau_2, \cdots, \tau_m$ of X.

If each $x \in X$ has a neighborhood U_x which meets only the cells $\tau_1, \tau_2, \cdots, \tau_m$, then $C(U_x) \subset \bigcup_i C(\tau_i)$ is a compact neighborhood of x.

Suppose that X is locally compact and $\sigma \subset X$. For each $x \in \sigma$, choose a compact neighborhood V_x of x, and then choose finitely many V_x, say $V_1, V_2, \cdots, V_n$ which cover the compact set σ.

Then σ meets at most the cells of $\bigcup_i C(V_i)$, which is a finite complex.

The proof for the countable case follows by replacing finite by countable in the proof above. ∎

LEMMA 3.7. *A closed cell of a CW complex is a metrizable subset.*

Proof. Let σ be a closed cell of X and choose a characteristic map $\varphi: D^n \to \sigma$, where D^n has the standard metric d. Define a metric $\bar{d}$ on σ by the formula $\bar{d}(x,y) = d(\varphi^{-1}(x), \varphi^{-1}(y))$. Since $\varphi^{-1}(x)$ is compact in D^n, this is always well defined. That this is a metric is easy to verify, the most difficult point being $d(x,y) = 0$ implies $x = y$. But since $\varphi^{-1}(x)$ and $\varphi^{-1}(y)$ are compact, if $d(\varphi^{-1}(x), \varphi^{-1}(y)) = 0$, then they must intersect, and thus $x = y$. The metric topology induced on σ by $\bar{d}$ is the quotient topology since φ is a distance decreasing function from the compact space D^n onto σ, which is a Hausdorff space with the $\bar{d}$ metric topology. ∎

PROPOSITION 3.8. *A connected CW complex X is metrizable if and only if it is locally finite.*

Proof. If X is locally finite and connected, it is countable by proposition I.3.18, and is therefore a countable union of subspaces with countable bases, namely, its cells. In the next section we will prove a CW complex is a normal space, and therefore X is metrizable.

Conversely, suppose that X is not locally finite and let $x \in X$ be a point that belongs to infinitely many cells. Then there cannot be a countable basis for the neighborhoods of x. For suppose that $\{V_i\}$ is a countable neighborhood base of x. We select a sequence $\{x_i\}$ such that $x_i \in V_i$ and each x_i lies in a distinct open cell. The set of x_i is closed because it meets each cell in a finite subset, hence cannot converge to x. But if V_i were a neighborhood basis for x, the sequence $\{x_i\}$ would have to converge to x. ∎

PROPOSITION 3.9. *A CW complex is connected if and only if it is connected in the carrier topology.*

Proof. If X is connected in the CW topology, it is connected in any coarser topology, and since the CW topology refines the carrier topology, X is connected in the carrier topology. Conversely, if X is connected in the carrier topology, then by proposition I.3.15, there is a c-path between any two points x,x' of X, i..e, a finite sequence of intersecting cells such that the first cell contains x and the last x'. Each cell is the continuous image of a connected set, hence is connected, and the union of such a sequence of intersecting cells must be connected. By the usual argument, X is connected. ∎

COROLLARY **3.10.** *Each component of a CW complex is a subcomplex.* ∎

PROPOSITION **3.11.** *A CW complex X is connected if and only if its 1-skeleton X^1 is connected.*

Proof. By proposition I.3.17, if X^1 is connected, X is carrier connected, hence is connected. Suppose that X^1 is not connected. Then $X^1 = A_1^1 \cup A_2^1$ where the A_i^1 are disjoint subcomplexes. We will prove inductively that the existence of such a partition of X^1 implies that there is a similar partition of each X^k. Thus suppose that $X^k = A_1^k \cup A_2^k$, where the A_i^k are disjoint subcomplexes and $k \geq 1$. If σ^{k+1} is a $(k+1)$-cell of X, then $\dot{\sigma}^{k+1} \subset X^k$ is connected, hence lies in one of the complexes A_i^k. Define

$$A_i^k \cup \mathbf{U}\{\sigma^{k+1} \mid \dot{\sigma}^{k+1} \subset A_i^k\} = A_1^{k+1}.$$

Then each A_1^{k+1} is a subcomplex of X^{k+1} and the A_1^{k+1} partition X^{k+1}. Thus if X^1 is not connected and we set $A_i = \mathbf{U}_k A_i^k$, the complexes A_i partition X. ∎

4. PARACOMPACTNESS

We refer the reader to Appendix I for a general discussion of paracompactness. We prove a preliminary lemma before the main theorem.

Lemma 4.1. *Let K be compact, let $C\subset K$ be closed, let $\mathfrak{U}=\{U_\alpha \mid \alpha\in A\}$ be an open covering of K, and let $\{p_\alpha \mid \alpha\in A\}$ be a partition of unity subordinated to $\mathfrak{U}\mid C=\{U_\alpha\cap C \mid U_\alpha\in\mathfrak{U}\}$. Then there is a partition of unity $\{\bar{p}_\alpha \mid \alpha\in A\}$ subordinated to $\mathfrak{U}$ such that for each $\alpha\in A$, $\bar{p}_\alpha \mid C=p_\alpha$.*

Proof. Let W_α denote the support of p_α, this being the set on which p_α takes nonzero values. First observe that since C is compact only finitely many functions p_α are nonzero because each $x\in C$ has a neighborhood which meets only finitely many sets W_α, and the covering $\{U_\alpha\}$ has a finite subcover. For each $\alpha\in A$ such that $W_\alpha\neq\varnothing$, choose an open set V_α in K such that $\bar{W}_\alpha\subset V_\alpha\subset \bar{V}_\alpha\subset U_\alpha$. Use Tietze's theorem to extend p_α to a map $p_\alpha':K\rightarrow I$ such that $p_\alpha'(K-V_\alpha)=0$. The family $\{p_\alpha'\}$ has the property that for some open V such that $C\subset V\subset \bigcup_\alpha V_\alpha$, the sum $\sum_\alpha p_\alpha'(x)$ is positive for $x\in V$. Finally, take any finite subcover

$$\{U_1,U_2,\cdots,U_n\}\subset\mathfrak{U}$$

and any subordinated partition of unity $\{q_1,q_2,\cdots,q_n\}$, and let $u:K\rightarrow I$ be an Urysohn function such that $u(C)=0$ and $u(K-V)=1$. Then $0<\sum_\alpha p_\alpha'(x)+\sum_i u(x)q_i(x)=q(x)$, and the restrictions of the functions p_α', q_1, q_2, $\cdots$, q_n to C give the family $\{p_\alpha'/q \mid \alpha\in A\}\cup\{q_1/q, q_2/q,\cdots,q_n/q\}$ which is a partition of unity subordinated to $\mathfrak{U}$ with the required property. ▮

Theorem 4.2. *A CW complex X is paracompact and is therefore a normal space.*

Proof. Let $\mathfrak{U}=\{U_\alpha \mid \alpha\in A\}$ be an open covering of the CW complex X. Since X is a Hausdorff space, we only need to show that there is a partition of unity subordinated to $\mathfrak{U}$.

Let the finite subcomplexes of X be indexed by Γ and for each subset $\Delta\subset\Gamma$ let $X_\Delta=\bigcup\{X_\gamma \mid \gamma\in\Delta\}$ and

$$\mathfrak{U}_\Delta=\mathfrak{U}\mid X_\Delta=\{U_\alpha\cap X_\Delta \mid \alpha\in A\}.$$

Let T denote the set of triples $(X_\Delta,\mathfrak{U}_\Delta,p)$, where p is a partition of unity subordinated to $\mathfrak{U}_\Delta$. The set T is nonempty, because we may take X_γ to be a vertex and p to be the function with range 1. When T is partially ordered in the obvious way by inclusion and restriction, any linearly ordered subset of T has an upper bound consisting of the union of the subcomplexes, the restricted covering, and the union of the partitions of unity. By Zorn's lemma there is a maximal triple $(X_{\Gamma_0}, U_{\Gamma_0}, p_0)$. If $X_{\Gamma_0} \neq X$, then for $x \in X - X_{\Gamma_0}$, $C(x)$ is not a subcomplex of X_{Γ_0}. The subset $X_{\Gamma_0} \cap C(x)$ is a closed subset of the compact set $C(x)$ and, by the lemma, $p \mid (C(x) \cap X_{\Gamma_0})$ extends over $C(x)$, hence over $X_{\Gamma_0} \cup C(x)$ which contradicts maximality of $(X_{\Gamma_0}, U_{\Gamma_0}, p_0)$. ∎

With a little more effort, we can also prove that each subset of a CW complex is paracompact.

PROPOSITION **4.3.** *A CW complex is perfectly normal (see Appendix I).*

Proof. By lemma 3.7 each cell of a CW complex is a metrizable space and therefore perfectly normal. A finite union of closed perfectly normal subsets of a space X is perfectly normal. Suppose C is closed in the union $\bigcup D_n$, where each D_n is perfectly normal and closed in X. First let f_1 be any nonnegative continuous function on D_1 zero exactly on $C \cap D_1$. If f_k is a nonnegative extension over $\bigcup\{D_i \mid i<k\}$ which is zero exactly on $C \cap \bigcup\{D_i \mid i \leq k\}$, extend $f_k \mid (D_{k+1} \cup \bigcap\{D_i \mid i \leq k\})$ to a function f_k'; $D_{k+1} \to I$ such that $f_k'(C \cap D_{k+1}) = 0$ by using Tietze's theorem. Since D_{k+1} is perfectly normal, there is a function v on D_{k+1} which vanishes exactly on $(C \cup \bigcup\{D_i \mid i \leq k\}) \cap D_{k+1}$. Then $f_{k+1} = f_k' + v$ extends f_k over D_{k+1} nonnegatively and vanishes only on the points of C. This process of extension finally yields a continuous function of the entire union vanishing only on C, and therefore the union is perfectly normal. In particular, a finite CW complex is perfectly normal.

For the general case, let X by any CW complex and C any closed subset. As in theorem 4.2, index the finite subcomplexes of X by Γ, and consider the collection of pairs (X_Δ, f_Δ), where $X_\Delta = \bigcup\{X_\gamma \mid \gamma \in \Delta\}$ and f_Δ is a continuous nonnegative function on X_Δ which vanishes exactly on $C \cap X_\Delta$. Any linearly ordered collection of such pairs has as upper bound the union of the subcomplexes and the union of the functions; hence, by Zorn's lemma, there is a maximal such pair (X^*, f^*). If $X^* \neq X$, then for $x \in X - X^*$, $C(x)$ is not a subcomplex of X^*, and an argument similar to the one above would extend $f^* \mid (X^* \cap C(x))$ over $C(x)$ so as to vanish exactly on the points of $C \cap C(x)$, and finally over $X^* \cup C(x)$. Thus X^* would not be maximal. ∎

COROLLARY **4.4.** *Every subset of a CW complex is paracompact and perfectly normal.*

Proof. Every subset of a space which is paracompact and perfectly normal is also paracompact and perfectly normal (see Appendix I, theorem 6). ∎

5. PRODUCTS, QUOTIENTS, AND ADJUNCTIONS

We wish to examine conditions under which the usual topological constructions on CW complexes lead to CW complexes. Sections 5, 6, and 7 of chapter I show that one may perform the constructions with the cell complexes, and thus the problem is to make sure the topologies work right.

If X_1 and X_2 are CW complexes, the product complex is a Hausdorff space, is closure finite, and has continuous characteristic maps. Let $X_1 \otimes X_2$ denote the space for which the underlying set is the product $X_1 \times X_2$ and the topology is the weak topology with respect to the product cells. Then $X_1 \otimes X_2$ is a CW complex. We will use $X_1 \times X_2$ to denote the product space. The following give the relation between these spaces with the same underlying set.

PROPOSITION **5.1.** *There is a continuous bijection* $\eta: X_1 \otimes X_2 \to X_1 \times X_2$. *Thus the spaces* $X_1 \otimes X_2$ *and* $X_1 \times X_2$ *differ only in that the CW topology may be finer than the product topology.*

Proof. Let $\pi_i: X_1 \otimes X_2 \to X_i$ be the projection maps for $i = 1, 2$.

If $\sigma \times \tau$ is a cell of $X_1 \otimes X_2$, then $\pi_i \mid (\sigma \times \tau)$ is clearly continuous, and by proposition 1.3, π_i is continuous for $i = 1, 2$. Thus the map $\eta = \pi_1 \times \pi_2 : X_1 \otimes X_2 \rightarrow X_1 \times X_2$ is a continuous bijection. ∎

THEOREM **5.2.** *Let X and Y be CW complexes. If either:*
(1) X (or Y) is finite; or
(2) both X and Y are countable;
then the map $\eta : X \otimes Y \rightarrow X \times Y$ is a homeomorphism. Thus the CW topology and the product topology on the product set $X \times Y$ coincide.

Proof. We first remark that if K and L are compact subsets of X and Y, respectively, and if W is an open subset of $X \times Y$ such that $K \times L \subset W$ there exist open sets $U \subset X$ and $V \subset Y$ such that $K \times L \subset U \times V \subset \bar{U} \times \bar{V} \subset W$. This is a modification of a theorem of Wallace (see Kelley [22, p. 142]) using the normality of X and Y.

To prove the theorem we only need to prove that a set W which is open in $X \otimes Y$ is also open in $X \times Y$. Let $(x_0, y_0) \in W$, and suppose that x_0 is interior to the cell σ of X and y_0 is interior to the cell τ of Y.

Suppose X is a finite complex. The set $B_{y_0} = \{x \mid (x, y_0) \in W\}$ is open and contains x_0, and since the set $X \times \{y_0\} \subset X \otimes Y$ is homeomorphic to the compact space X, it contains a compact neighborhood $\bar{U}$ of x_0. Let $V = \{y \mid \bar{U} \times \{y\} \subset W\}$. By the remark above, this set meets each cell carrier $C(\tau)$ containing y_0 in a neighborhood of y_0, and by proposition 3.3 is a neighborhood of y_0. The set $\bar{U} \times V = \bigcup \{\bar{U} \times \{y\} \mid y \in V\}$ is thus a neighborhood of (x_0, y_0) in the product topology and is contained in W. Thus W is open in the product topology.

Suppose that X and Y are countable. Order the cells σ_i, τ_i of X and Y, respectively, so that $x_0 \in \sigma_1 - \dot{\sigma}_1$ and $y_0 \in \tau_1 - \dot{\tau}_1$. Let $A_k = \bigcup_{i=1}^{k} C(\sigma_i)$ and define B_k similarly. Then A_k and B_k are finite subcomplexes of X and Y, respectively, for $i = 1, 2, \cdots$, and $X = \bigcup_k A_k$, $Y = \bigcup_k B_k$. By the first part of the theorem, for each integer k, $A_k \otimes B_k$ has the product topology. Choose open sets $U_i \subset A_i$ and $V_i \subset B_i$ such that $(x_0, y_0) \in U_i \times V_i \subset U_{i+1} \times V_{i+1} \subset W \cap (A_{i+1} \times B_{i+1})$. This may be done because U_i and V_i are compact. Now let $U = \bigcup_i U_i$ and $V = \bigcup_i V_i$. Then $(x_0, y_0) \in$

$U \times V \subset W$, and for each i, $U \cap A_i$ and $V \cap B_i$ contain neighborhoods of x_0 and y_0, respectively. Thus by proposition 3.3, U and V are neighborhoods of x_0 and y_0, respectively. ∎

COROLLARY **5.3.** *A subset A of $X \otimes Y$ is compact if and only if $\eta(A)$ is compact in $X \times Y$.*

Proof. Since η is continuous, the compactness of A implies that of $\eta(A)$. If $\eta(A)$ is compact, then its projections A_1 and A_2 in X and Y are images of a compact set under continuous maps and so are compact. Thus they have finite (and compact) carriers $C(A_1)$ and $C(A_2)$, and A lies in the finite subcomplex $C = C(A_1) \times C(A_2)$ of $X \otimes Y$. The map η is a homeomorphism on C by 5.2, so that A is compact in C, and therefore in $X \otimes Y$. ∎

COROLLARY **5.4.** *If X is a CW complex and I is the CW complex on the unit interval I consisting of two 0-cells and a 1-cell, then $X \times I$ is a CW complex. Consequently, a map $H: X \times I \to Y$ is continuous if and only if $H \mid \sigma \times I$ is continuous for each cell $\sigma \subset X$.*

Proof. The first part is a direct application of the theorem. The statement about continuity follows from proposition 1.3, if we observe that each cell of $X \times I$ is a subset of a cell $\sigma \times I$. ∎

COROLLARY **5.5.** *Let X and Y be CW complexes. If either:*
(1) X is locally compact, or
(2) both X and Y are locally countable,
then $X \otimes Y$ and $X \times Y$ are homeomorphic.

Proof. The proof is by the usual localization technique. Let $(x,y) \in X \times Y$, and suppose $(x,y) \in W$ which is open in $X \times Y$. Let K be a compact neighborhood of x. By the theorem, there exist sets U and V open in $C(K)$ and Y, respectively, such that $(x,y) \in U \times V \subset W \cap (C(K) \times Y)$. But K is a neighborhood of x, so that $U \cap K$ is a neighborhood of x, and $(x,y) \in (U \cap K) \times V \subset W$.

The proof of (2) is similar and is omitted. ∎

The following example due to Dowker [6] shows that when neither (1) nor (2) of the corollary is satisfied, the space $X \otimes Y$ may have a topology finer than the product topology.

Let X be the cone over a discrete countably infinite set with the CW topology, and let Y be the cone over a discrete uncountably infinite set with the CW topology. Then $X \otimes Y$ has a topology finer than the product topology. The details will be omitted.

The construction of quotient complexes involves the usual problems about quotient spaces satisfying the Hausdorff property. In this connection, the following is useful for our purposes. The proof will be omitted.

Lemma **5.6.** *If X is a normal space and if the quotient map $X \to X/\mathcal{R}$ is a closed map, then $X/\mathcal{R}$ with the quotient topology is a normal space.* ∎

Proposition **5.7.** *Let X be a CW complex and $\mathcal{R}$ a cellular equivalence relation such that the quotient space $X/\mathcal{R}$ is Hausdorff. Then with the quotient structure on $X/\mathcal{R}$, the quotient space is a CW complex and the quotient map is a regular cellular map.*

Proof. The quotient structure on the quotient set is the quotient of a closure finite structure, and therefore by proposition I.6.7, is itself closure finite. The quotient map, as a map of cell complexes, is known from proposition I.6.4 to be a regular cellular map. Therefore we have only to prove that $X/\mathcal{R}$ is a CW complex to finish the proof.

Since by hypothesis the quotient space is Hausdorff, and since the characteristic maps for a cell of the quotient structure are compositions of characteristic maps for X with the quotient map, $X/\mathcal{R}$ is Hausdorff and the characteristic maps of its structure are continuous. Hence we only have to prove that the quotient topology is the weak topology with respect to the cells of the quotient structure. Let C be any subset of $X/\mathcal{R}$ such that $p^{-1}(C)$ is closed in X, where p is the quotient map. Then for any cell $\bar{\sigma}$ of $X/\mathcal{R}$ and corresponding $\mathcal{R}$-minimal cell σ in X, $\sigma \cap p^{-1}(C) =$

$(p \mid \sigma)^{-1}(C)$ is closed and therefore compact. Thus $p(p^{-1}(C)) \cap \bar{\sigma}$ is closed. Conversely, if $C \cap \bar{\sigma}$ is closed for each cell $\bar{\sigma}$ of $X/\mathcal{R}$, then for any cell τ mapped by p onto $\bar{\sigma}$, the map $p \mid \tau$ is a continuous map of a compact space onto a Hausdorff space $\bar{\sigma}$, hence is a quotient map, and $p^{-1}(C) \cap \tau = p^{-1}(C \cap \bar{\sigma}) \cap \tau$ is closed. Therefore, if C meets each cell of $X/\mathcal{R}$ in a closed set, $p^{-1}(C)$ meets each cell of X in a closed set, and C is closed in the quotient topology. ∎

One may check that the cellular equivalence relation described in I.6, collapsing subcomplexes to vertices, satisfies the conditions of corollary 5.5 when X is a CW complex. Thus the quotient complex is a CW complex. In particular, the cones and suspensions of CW complexes are CW complexes, and if the product of CW complexes is a CW complex, the smash product is also.

The following is a noteworthy special case of proposition 5.7.

PROPOSITION **5.8.** *If $\mathcal{R}$ is a cellular equivalence relation on the CW complex X and if for each cell τ associated with the $\mathcal{R}$-minimal cell σ, the quotient map $p_\tau : \tau \to p(\sigma)$ factors as $p_\tau = p_\sigma f_{\sigma,\tau}$ for some continuous map $f_{\sigma,\tau} : \sigma \to \tau$, then $X/\mathcal{R}$ is a CW complex.*

Proof. Case (1): finite CW complexes. We will prove by induction that the quotient map $p : X \to X/\mathcal{R}$ is a closed quotient map. This is clearly true for 0-dimensional complexes. We assume it is true for complexes of dimension less than k, and assume that X is a complex of dimension k. Let C be closed in X and suppose that C is contained in the k-cell τ. If τ is not $\mathcal{R}$-minimal, let σ be a corresponding $\mathcal{R}$-minimal cell. The cell σ must have dimension less than k. Since $p_\tau = p_\sigma f_{\sigma,\tau}$, if $f = f_{\sigma,\tau}$, then $p_\tau^{-1} p_\tau(C) = f^{-1} p_\sigma^{-1} p_\sigma f(C) = f^{-1} f(C)$ because σ is $\mathcal{R}$-minimal. Since f is a continuous map from the compact space τ to the Hausdorff space σ, it follows that $p_\tau^{-1} p_\tau(C) = f^{-1} f(C)$ is closed. If τ is $\mathcal{R}$-minimal, $p_\tau^{-1} p_\tau(C) = C \cup p_\tau^{-1} p_\tau(C \cap \dot{\tau})$. By the induction hypothesis, $p_\tau^{-1} p_\tau(C \cap \dot{\tau})$ is closed, and therefore $p_\tau^{-1} p_\tau(C)$ is a union of two closed sets and is closed. Thus, in either case, $p_\tau(C)$ is closed in the quotient topology. Since X is covered by a finite number of cells of dimension no greater than k, and on each p restricts to a closed continuous map, it follows that p is closed on all of X. Thus by lemma 5.6 and proposition 5.7, $X/\mathcal{R}$ is a CW complex.

Case (2): the general case. Since the quotient map $p:X\to X/\mathcal{R}$ is closed and continuous in the carrier topology by I.6.6, $p(C(\sigma))=C(p(\sigma))$. Thus each cell carrier $C(p(\sigma))$ in $X/\mathcal{R}$ is a CW complex by case (1). Also, any subcomplex of the cell complex on $X/\mathcal{R}$ is closed in the quotient topology because its preimage is a subcomplex of X, hence is a closed set. Let K be a subset of $X/\mathcal{R}$. If $p^{-1}(K)$ is closed in X, then it meets each carrier $C(\sigma)$ in a closed set, and since $p \mid C(\sigma)$ is a closed map, $K\cap C(p(\sigma))$ is closed in $C(p(\sigma))$ for each cell carrier $C(p(\sigma))$ of $X/\mathcal{R}$. Thus a set closed in the quotient topology is closed in the weak topology with respect to the cell carriers. If $K\cap C(p(\sigma))$ is closed for each σ, then $p^{-1}(K)\cap C(\sigma) = (p \mid C(\sigma))^{-1}(K\cap C(p(\sigma)))$ is closed in $C(\sigma)$ by case (1). By corollary 3.5, the topology on X is dominated by the cell carriers of X, and we conclude that $p^{-1}(K)$ is closed in X. Thus K is closed in the quotient topology. Since the weak topology on $X/\mathcal{R}$ with respect to the cell carriers is the same as the quotient topology, by another application of corollary 3.5, $X/\mathcal{R}$ is a CW complex. ∎

If X is a CW complex, a *family of identifications* Ω on X is a family of identifications in the sense of I.6, in which each $\omega\in\Omega$ is a regular cellular homeomorphism. We abbreviate the quotient complex $X/\mathcal{R}_\Omega$ to X/Ω.

Corollary 5.9. *If Ω is a family of identifications on the CW complex X, then X/Ω is a CW complex.* ∎

In order to study adjunction complexes, we need the following fact about the adjunction sets.

Lemma 5.10. *Let X, Y, Z, be sets, let $A\subset X$, $B\subset Y$, and let $f:A\to Y$ and $g:B\to Z$ be functions. Suppose $p:Z\cup Y\to Z\cup_g Y$ and $q:Y\cup X\to Y\cup_f X$ are the quotient maps. Then as quotients of the disjoint* *union* $Z\cup Y\cup X$, $(Z\cup_g Y)\cup_{pf} X = Z\cup_{\bar g}(Y\cup_f X)$, *where* $\bar g = g(q \mid B)^{-1}$.

Proof. One checks that the equivalence classes of $w\in Z\cup Y\cup X$ are the same under the composite maps

$$Z\cup Y\cup X \xrightarrow{p\cup 1_X} (Z\cup_g Y)\cup X \to (Z\cup_g Y)\cup_{pf} X$$

and

$$Z\cup Y\cup X \xrightarrow{1_X\cup q} Z\cup(Y\cup_f X) \to Z\cup_{\bar g}(Y\cup_f X).$$

We omit the details. ∎

THEOREM **5.11.** *Let X and Y be CW complexes, let $A\subset X$ be a subcomplex, and let $f: A\to Y$ be a continuous cellular map. Then the adjunction space $Y\cup_f X$ is a CW complex.*

Proof. By proposition I.7.4, $Y\cup_f X$ is a closure finite cell complex. Since the quotient map $p: Y\cup X\to Y\cup_f X$ is continuous, the characteristic maps of $Y\cup_f X$ are continuous. We need only check that $Y\cup_f X$ is Hausdorff and has the weak topology with respect to its cells.

Note that $Y\cup_f (A\cup X^0)$ is a disjoint union of a CW complex with a discrete set, hence is a CW complex. Suppose that $Y\cup_f (A\cup X^k)$ is a CW complex for $k<n$. Then if $\mathcal{E}^n = \mathbf{U}\{E_\sigma^n \mid \sigma$ is an n-cell and $\sigma - \dot\sigma \subset X-A\}$ is the disjoint union and $\dot F: \dot{\mathcal{E}}^n \to X^{n-1}$ is the restriction of the union of the corresponding characteristic maps, by the lemma we have

$$\begin{aligned} Y\cup_f(A\cup X^n) &= Y\cup_f((A\cup X^{n-1})\cup_{\dot F}\mathcal{E}^n) \\ &= (Y\cup_f(A\cup X^{n-1}))\cup_{p\dot F}\mathcal{E}^n. \end{aligned}$$

By hypothesis $Y\cup_f (A\cup X^{n-1})$ is a CW complex, and by proposition 2.2, if we attach n-cells properly, as we have done here, the adjunction is a CW complex. By induction we conclude that

$Y\cup_f(A\cup X^n)$ is a CW complex for all n. One easily checks that $Y\cup_f(A\cup X^n)$ is a closed subset of $Y\cup_f X$ for each n, and that $Y\cup_f X=\bigcup_n Y\cup_f(A\cup X^n)$.

If $C\cap(Y\cup_f(A\cup X^n))$ is closed for each n, then $p^{-1}(C)\cap(Y\cup(A\cup X^n))$ is closed for each n, so that $p^{-1}(C)$ is closed by proposition 3.3. Since $Y\cup_f X$ has the quotient topology, C is closed in $Y\cup_f X$. If C is closed in $Y\cup_f X$, clearly $C\cap(Y\cup_f(A\cup X^n))$ is closed for each n. Thus $Y\cup_f X$ has the weak topology with respect to the $Y\cup_f(A\cup X^n)$ and by a trivial modification of theorem 2.4, $Y\cup_f X$ is a CW complex. ∎

Corollary **5.12.** *The following are CW complexes:*

(1) *the mapping cylinder of a continuous cellular map between CW complexes;*

(2) *the join of two CW complexes whenever their product is a CW complex;*

(3) *the cone and reduced cone over a CW complex;*

(4) *the suspension and reduced suspension of a CW complex;*

(5) *the smash product of two CW complexes whenever their product is a CW complex.*

Proof. These follow from the discussion in I.7 and the theorem above. ∎

6. *HOMOTOPY AND LOCAL PROPERTIES*

We recall (see 0.3) that a subspace $A\subset X$ is a strong deformation retract provided there is a continuous map $\rho:X\times I\to X$ such that $\rho(x,0)=x$, $\rho(x,1)\in A$, and $\rho(a,t)=a$ for $a\in A$.

Theorem **6.1.** *Let A be a subcomplex of a CW complex X and let V be an open neighborhood of A. Then there is an open set $U\subset X$ such that $A\subset U\subset V$ and A is a strong deformation retract of U.*

Proof. Case (1): finite CW complexes. We will construct U and a strong deformation retraction of U onto A inductively over the skeletons of X as a subset of the union of the carriers of cells

$\sigma \subset X$ such that $(\sigma - \dot{\sigma}) \cap V \neq \emptyset$. Choose $U^0 = A$ and ρ^0 to be the map $\rho^0(a,t) = a$.

Suppose that for all $k < n$ we have the following data:

(1) a set U^k which is open in $A \cup X^k$ such that $A \subset U^k$ and $\bar{U}^k \subset V^k = V \cap (A \cup X^k)$, where $\bar{U}^k$ is the closure of U^k, and such that $U^k \cap (A \cup X^{k-1}) = U^{k-1}$;

(2) a strong deformation retraction ρ^k of U^k onto A such that $\rho^k(x,t) = \rho^k(x,0)$ for $0 \leq t \leq 2^{-k}$ and such that $\rho^k \mid U^{k-1} \times I = \rho^{k-1}$. We will extend these to a set U^n and a strong deformation retraction ρ^n of U^n onto A which has these same properties.

For each n-cell σ_λ whose interior meets V, choose a characteristic map $\varphi_\lambda: D^n \to \sigma_\lambda \subset X$, where D^n is the unit n-disc in R^n. Since V is open, $V \cap \sigma_\lambda$ is open in σ_λ, and $\varphi_\lambda^{-1}(V) = V_\lambda{}^n$ is open in D^n. Moreover, $\varphi_\lambda^{-1}(U^{n-1}) \subset \varphi_\lambda^{-1}(V^{n-1}) = \varphi_\lambda^{-1}(V^n) \cap \dot{D}^n$, and $\overline{\varphi_\lambda^{-1}(U^{n-1})} \subset \varphi_\lambda^{-1}(\bar{U}^{n-1}) \subset V_\lambda{}^n \cap \dot{D}^n$. Now $\overline{\varphi_\lambda^{-1}(U^{n-1})}$ and $D^n - V_\lambda{}^n$ are disjoint compact sets in the metric space D^n, hence are a positive distance $4d_\lambda$ apart unless $\varphi_\lambda^{-1}(U^{n-1}) = \emptyset$. Let $W_\lambda{}^n = \{x \in D^n \mid x = (1-s)z,\ z \in \varphi_\lambda^{-1}(U^{n-1}),$ and $0 \leq s < d_\lambda\}$ for $\varphi_\lambda^{-1}(U^{n-1}) \neq \emptyset$, and $W_\lambda{}^n = \emptyset$ otherwise. Then $W_\lambda{}^n \subset \bar{W}_\lambda{}^n \subset V_\lambda{}^n$, $W_\lambda{}^n$ can be deformed radially onto $\varphi_\lambda^{-1}(U^{n-1})$ (since $(0,0\cdots 0) \notin \bar{W}_\lambda{}^n$). For example, for our purposes it is convenient to choose $r_\lambda{}^n: W_\lambda{}^n \times [0, 2^{-n+1}] \to W_\lambda{}^n$ as follows: for $0 \leq t \leq 2^{-n}$, $r_\lambda{}^n(x,t) = x$, and for $2^{-n} \leq t \leq 2^{-n+1}$, $r_\lambda{}^n(x,t) = (2 - 2^n t)x + (2^n t - 1)(x / \langle x,x \rangle^{1/2})$.

Let $U_\lambda{}^n = \varphi_\lambda(W_\lambda{}^n)$, and define $U^n = U^{n-1} \cup (\mathbf{U}_\lambda U_\lambda{}^n)$. Then $U^n \subset \bar{U}^n \subset V^n$ and U^n is open in $A \cup X^n$ because $U^n \cap (A \cup X^{n-1}) = U^{n-1}$ is open in X^{n-1}, and $U^n \cap \sigma_\lambda$ is open in σ_λ since $\varphi_\lambda^{-1}(U^n) = W_\lambda{}^n$ which is open in D^n and σ_λ has the quotient topology with respect to φ_λ. Define the map ρ^n as follows: $\rho^n(x,t) = \varphi_\lambda r_\lambda{}^n(\varphi_\lambda^{-1}(x),t)$ for $0 \leq t \leq 2^{-n+1}$ and $x \in \sigma_\lambda$, and $\rho^n(x,t) = \rho^{n-1}(\rho^n(x, 2^{-n+1}), t)$ for $2^{-n+1} \leq t \leq 1$. Then ρ^n is a strong deformation retraction of U^n onto A, $\rho^n \mid U^{n-1} \times I = \rho^{n-1}$, and $\rho^n(x,t) = \rho^n(x,0)$ for $0 \leq t \leq 2^{-n}$.

Case (2): the general case. Fix a collection of characteristic maps for X. Given any open neighborhood V of A, we will construct a neighborhood U contained in $\overline{\text{st}}(A)$ (see I.3.19) and a strong deformation retraction of U onto A. Since $\overline{\text{st}}(A)$ is a closed neighborhood of A, we may assume that $X = \overline{\text{st}}(A)$ so that each cell carrier meets A.

In the first part of the proof, we constructed for each cell carrier $C(\sigma)$ a relatively open set U_σ containing $A \cap C(\sigma)$ and a strong deformation retraction ρ_σ of U_σ onto $A \cap C(\sigma)$. Examination of this construction shows that for each cell τ of $C(\sigma)$, $U_\sigma \cap \tau$ depends only on the dimension of τ, the construction of U_σ on the cells of the carrier of τ, and $V \cap \tau$. Hence on the intersection $C(\sigma) \cap C(\tau)$ of carriers, $U_\sigma \cap C(\sigma) \cap C(\tau) = U_\tau \cap C(\sigma) \cap C(\tau)$, and ρ_σ and ρ_τ restrict to the same strong deformation retraction of this open neighborhood onto $A \cap C(\sigma) \cap C(\tau)$. Thus the set $U = \mathbf{U}_\sigma U_\sigma$ is a set such that $A \subset U \subset V$, and $\rho = \mathbf{U}_\sigma \rho_\sigma$ is a function. Since the cell carriers dominate X, the set U is open and the function ρ is continuous, hence a strong deformation retraction of U onto A.

To complete the proof, set $U = \mathbf{U}_n U^n$ and $\rho = \mathbf{U}_n \rho^n$. Then $U \subset \bar{U} \subset V$ and ρ is a strong deformation retraction of U onto A. ∎

Since each vertex of a CW complex is a subcomplex, each vertex is a strong deformation retract of some neighborhood. The following results will enable us to prove that this is true of all points of a CW complex.

Definition **6.2.** Let X be a CW complex with respect to two families of cells $\mathcal{S}$ and $\mathcal{S}'$. Then $(X,\mathcal{S}')$ is a subdivision of $(X,\mathcal{S})$ provided for each $\sigma' \in \mathcal{S}'$ there is a cell $\sigma \in \mathcal{S}$ such that $\sigma' - \dot{\sigma}' \subset \sigma - \dot{\sigma}$.

Proposition **6.3.** *Let X be a CW complex with cells $\mathcal{S}$, let $A \subset X$ be a subcomplex with cells $\mathfrak{I}$, and let $(X,\mathcal{S}')$ be a subdivision of $(X,\mathcal{S})$. Then if $\mathfrak{I}' = \{\sigma' \in \mathcal{S}' \mid \sigma' \subset A\}$, $(A,\mathfrak{I}')$ is a subcomplex of $(X,\mathcal{S}')$, and is a subdivision of $(A,\mathfrak{I})$.*

Proof. Suppose that $\sigma' \in \mathcal{S}'$ and $(\sigma' - \dot{\sigma}') \cap A \neq \varnothing$. Then for some $\sigma \in \mathcal{S}$, $\sigma' - \dot{\sigma}' \subset \sigma - \dot{\sigma}$, and $(\sigma - \dot{\sigma}) \cap A \neq \varnothing$, so $\sigma' \subset \sigma \subset A$. Thus by proposition I.1.6, $(A,\mathfrak{I}')$ is a subcomplex of $(X,\mathcal{S}')$. The subdivision property is clearly inherited by the subcomplex $(A,\mathfrak{I}')$. ∎

Theorem **6.4.** *If X is a CW complex and $x \in X$, there is a subdivision of X in which x is a vertex.*

Proof. Let $\mathcal{S}$ be a set of cells for X, and let $x \in \sigma^n - \dot{\sigma}^n$. The proof is by induction on n. If $n=1$ or 0 there is little or nothing (respectively) to prove. Suppose that for each point $y \in X^{n-1}$ there is a subdivision $(X,\mathcal{S}_y')$ of $(X,\mathcal{S})$ such that y is a vertex of $\mathcal{S}_y'$. Let $y \in \dot{\sigma}^n$ be a vertex of $\mathcal{S}_y'$, let φ_σ be a characteristic map for σ^n, and choose $\theta,\bar{y} \in E^n = D^n$ so that $\varphi_\sigma(\theta) = x$ and $\varphi_\sigma(\bar{y}) = y$, where we assume $\theta = (0,0, \cdots, 0)$. Let $\varphi^0 : E^0 \to x \in X$, let $\varphi^1 : I \to X$ be $\varphi^1(t) = \varphi_\sigma(t\bar{y})$, and let $\tau^1 = \varphi^1(I)$. Choose a map $\psi : D^n \to D^n$ which collapses the lower half of the disc D^n onto the line segment between $\theta = (0,0, \cdots, 0)$ and $p = (0,0, \cdots, -1)$ by collapsing concentric hemispheres to points and maps the upper half of the disc

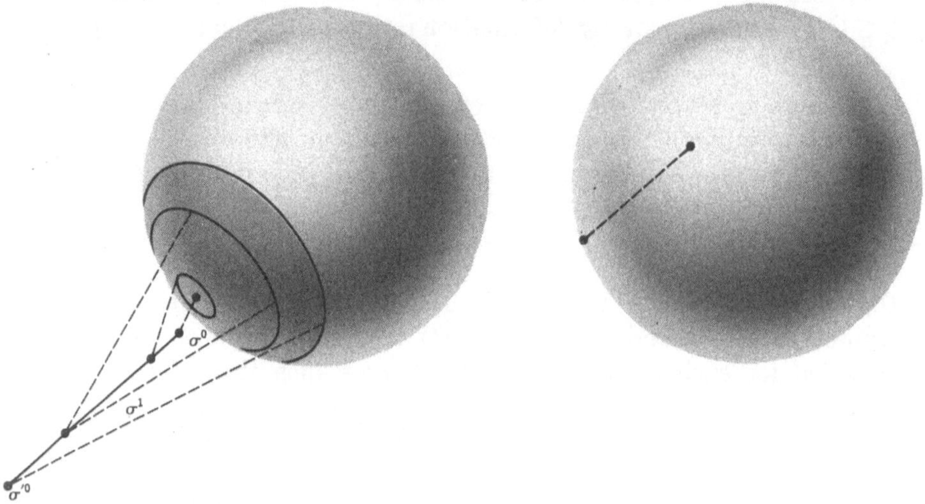

Figure 2.1. In the shaded portion of $\mathbf{S}^{n-1} = \dot{E}^n$ each concentric small sphere maps into the same point of I. After attaching by such a map, the corresponding n-cell "contains" the 1-cell σ^1.

homeomorphically onto the complement of this line segment (see Fig. 2.1). An analytic formula for ψ may be given as follows: First let $\psi' : \mathbf{S}^{n-1} \to \mathbf{S}^{n-1}$ be given as

$$\psi'(x_1,x_2, \cdots, x_n) = (2x_1x_n, 2x_2x_n, \cdots, 2x_{n-1}x_n, 1 - 2x_n^2)$$

for $x_n \geq 0$, and $\psi'(x_1,x_2, \cdots, x_n) = p$ for $x_n \leq 0$; and then using "polar coordinates" (t,z) in D^n, where $t \in I$ and $z \in \mathbf{S}^{n-1}$, let $\psi(t,z) =$

$t\psi'(z)$. Now define $\varphi^n : D^n \to X$ as $\varphi^n = \varphi_\sigma \psi$, and set $\tau^n = \varphi^n(D^n)$. Let $\mathcal{S}'' = (\mathcal{S}_y' - \{\sigma^n\}) \cup \{x, \tau^1, \tau^n\}$, i.e., we remove the cell σ^n and add in the new cells x, τ^1, and τ^n. It is easily checked that $(X, \mathcal{S}'')$ is a closure finite cell complex and X has the weak topology with respect to $\mathcal{S}''$. Clearly $(X, \mathcal{S}'')$ is a subdivision of $(X, \mathcal{S})$ in which x is a 0-cell. ∎

Definition **6.5.** A space X is *locally contractible at a point* $x \in X$ if each neighborhood V of x contains a neighborhood U of which x is a strong deformation retract. The space X is *locally contractible* if it is locally contractible at each point.

THEOREM **6.6.** *A CW complex is locally contractible.*

Proof. Each vertex of a CW complex is a subcomplex, and theorem 6.1 shows that a CW complex is locally contractible at each vertex. But by theorem 6.4 every point of a CW complex is a vertex in a suitable subdivision. ∎

COROLLARY **6.7.** *A CW complex is locally pathwise connected (and therefore locally connected).*

Proof. A contractible set is pathwise connected. ∎

COROLLARY **6.8.** *An open subset U of a CW complex is connected if and only if U is pathwise connected.*

Proof. This is true of any locally pathwise connected space. ∎

We should remark that the theorem of local contractibility can be proved without the subdivision theorem. Its proof is similar to the proof of theorem 6.1.

7. THE HOMOTOPY EXTENSION THEOREM

In the study of the set of homotopy classes of maps $f: X \to Y$, one often needs to know that a map defined and homotopic to f

on a subset of X can be extended over X. As we shall see, this is the case if X is a CW complex and the subset is a subcomplex.

LEMMA 7.1. *Let* (X,A) *be a pair such that* A *is closed in* X *and* $X\times I$ *is a normal space. If there is a neighborhood* U *of which* $(X\times\{0\})\cup(A\times I)$ *is a retract, then any map* $H':(X\times\{0\})\cup(A\times I)\to Y$ *extends over* $X\times I$.

Proof. Assume that U is open, let $\rho:U\to(X\times\{0\})\cup(A\times I)$ be a retraction, and let $u:X\times I\to I$ be an Urysohn function such that $u((X\times\{0\})\cup(A\times I))=1$ and $u((X\times I)-U)=0$. Then the map u' defined by $u'(x,t)=(x,t\cdot u(x,t))$ is a retraction of $X\times I$ onto U so that the composition $\rho u'$ retracts $X\times I$ onto $(X\times\{0\})\cup(A\times I)$. Thus the composition $H=H'\rho u'$ extends H' over $X\times I$. ∎

THEOREM 7.2. (*Homotopy Extension Theorem*). *Let* X *be a CW complex, and let* A *be a subcomplex. Then* (X,A) *has the homotopy extension property with respect to any space* Y (*see* 0.4).

Proof. Suppose that $f:X\to Y$ is a map and $h:A\times I\to Y$ is a homotopy such that $h(a,0)=f(a)$. Define $H':(X\times\{0\})\cup(A\times I)\to Y$ by $H'\mid(A\times I)=h$ and $H'(x,0)=f(x)$. By theorem 5.2, $X\times I$ is a CW complex, hence is normal, and by theorem 6.1, the subcomplex $(X\times\{0\})\cup(A\times I)$ has a neighborhood of which it is a (deformation) retract. By the lemma, H' extends to a map $H:X\times I\to Y$. ∎

COROLLARY 7.3. *If* (X,A) *is a CW pair, then* $i:A\to X$ *is a cofibration.* ∎

COROLLARY 7.4. *If* Y *is a CW complex and* $H:(A\times I)\cup(X\times\{0\})\to Y$ *is carried by the cellular carrier* E *of* $f=H\mid(X\times\{0\})$, *then there is an extension* $H:X\times I\to Y$ *which is carried by* E.

Proof. We proceed by induction on the dimension of skeletons. In dimension 0, if σ^0 is a cell of $X^0 - A$, then $C(f(\sigma^0)) = E(\sigma^0)$ is a subcomplex of Y which is path connected, so there is a path $H: \sigma^0 \times I \to Y$ which is carried by E. If we have an extension of H to $X^{n-1} \times I$ which is carried by E, and if σ is an n-cell of $X^n - A$, then by theorem 7.2, the map

$$H \mid ((((C(\sigma) \cap A) \cup C(\dot{\sigma})) \times I) \cup (C(\sigma) \times \{0\}))$$

may be extended to a map $H_\sigma : C(\sigma) \times I \to C(f(\sigma)) = E(\sigma)$. The maps H_σ fit together to give an extension over $X^n \times I$ carried by E. ∎

An argument similar to that in the proof of theorem 7.2 yields the following.

THEOREM **7.5.** *If (X,A) is a pair of CW n-ads, and if $H: (\mathbf{A} \times I) \cup (\mathbf{X} \times \{0\}) \to \mathbf{Y}$ is a map of n-ads, then there is an extension $H: \mathbf{X} \times I \to \mathbf{Y}$ which is a map of n-ads.* ∎

8. THE CELLULAR APPROXIMATION THEOREM

In the study of continuous maps between CW complexes, we would like to take advantage of the cellular structure of the two spaces. It turns out that, in general, the most useful maps between CW complexes which preserve "enough" structure are the continuous cellular maps. Following definition I.4.1, we say a map $f: X \to Y$ between the CW complexes X and Y is cellular if $f(X^n) \subset Y^n$ and f is continuous. These maps play a role in the theory of CW complexes analogous to that played by simplicial maps in the theory of simplicial complexes.

The proof of our main theorem will use some important facts about Euclidean space.

Definition **8.1.** The *order of a covering* $\mathcal{U}$ of a space is one more than the largest number of sets of $\mathcal{U}$ with a nonempty intersection. If no such maximum exists, the order is ∞. The *mesh* of a covering $\mathcal{U}$ of a metric space is the supremum (perhaps ∞) of the diameters of the sets of the covering.

LEMMA 8.2. *For any $\epsilon>0$, there is an open covering of the n-disc D^n of order no greater than n and mesh less than ϵ.*

(We remark that, in fact, one can prove by more advanced methods that there are not arbitrarily small coverings of order less than n; hence the order property of coverings may be used to give a topological definition of dimension. See Hurewicz and Wallman [20].)

Proof. It will suffice to give an open cover of R^n with bounded mesh M and order no greater than n. Then, by a dilitation by a positive factor less than ϵ/M, we obtain a cover of R^n of order no greater than n and mesh less than ϵ. The restriction of this cover to D^n is a cover with the desired properties.

The simplest way to come by such a cover is by a "staggered brick" tiling of R^n. One starts by constructing a closed cover, the elements of which are cubes of side one with sides parallel to the coordinate planes. The centers of these cubes will be located inductively as follows. Subdivide $R=R^1$ into closed unit intervals about its integer points. Suppose the centers of the cubes are at the points $(x_1^{(i)},x_2^{(i)},\cdots,x_k^{(i)})$ in a tiling of R^k. Let b be a transcendental number, and choose centers for a tiling of R^{k+1} to be the points $(x_1^{(i)}+b^k,x_2^{(i)}+b^k,\cdots,x_k^{(i)}+b^k,2m+1)$, and $(x_1^{(i)},x_2^{(i)},\cdots,x_k^{(i)},2m)$, where m is an integer.

Assume inductively that the order of the tiling of R^k described is no greater than k. This is clearly so if $k=1$. First note that if $\{C_i(a)\mid i=1,2,\cdots,r\}$ is a collection of tiling cubes of R^{k+1} which have centers with $(k+1)$-st coordinate a, then their intersection will be contained in an intersection of hyperplanes $x_i=c$, where $i\leq k$ and c is the coordinate of a k-dimensional edge of a cube in the collection. Such a number c will be a linear combination of $(1/2)$, b, b^2, $\cdots$, b^k with integer coefficients in which b^k has coefficient 0 if and only if a is even. We must investigate the intersection of two sets

$$C_1(a-1)\cap\cdots\cap C_r(a-1)=A \quad\text{and}\quad C_1(a)\cap\cdots\cap C_s(a)=B,$$

where a is an integer. By the inductive hypothesis, $r\leq k+1$ and

$s \leq k+1$. Let I be the set of induces i such that A is contained in the intersection of the coordinate hyperplanes $x_i = c$, and let J be the same for B. If $I \cap J \neq \emptyset$, the fact that b is transcendental implies that A and B lie in disjoint parallel affine subspaces of R^{k+1}. Thus we must have either $r=1$ or $s=1$, and at most $k+2$ cubes can meet. If $I \cap J = \emptyset$, the sets A and B lie in orthogonal $(k-r+1)$- and $(k-s+1)$-dimensional affine subspaces and their intersection has dimension $k-r-s+2$ at most. Thus if $A \cap B \neq \emptyset$, $k-r-s+2 \geq 0$, and $r+s \leq k+2$. ∎

LEMMA **8.3.** *Let $K \subset R^n$ be compact and let $\mathcal{U}$ be an open cover of K. Then there is a refinement of $\mathcal{U}$ which is of order n.*

Proof. Let ϵ be the Lebesgue number of the covering $\mathcal{U}$. Choose a covering $\mathcal{U}'$ of R^n of order n and mesh less than ϵ. Then the restriction $\mathcal{U}' \mid K = \{U' \cap K \mid U' \in \mathcal{U}'\}$ is of order n and refines $\mathcal{U}$. ∎

Using this fundamental lemma which describes the topological dimension of R^n, we prove the following result.

LEMMA **8.4.** *Let V be open in R^n with $\bar{V}$ compact. Let $g : \bar{V} \to D^m$ be a continuous map such that $g^{-1}(D^m - \dot{D}^m) \subset V$. Then if $n < m$, g is homotopic to a map g' relative to $\dot{V} = \bar{V} - V$ which omits an interior point of D^m* (Fig. 2.2).

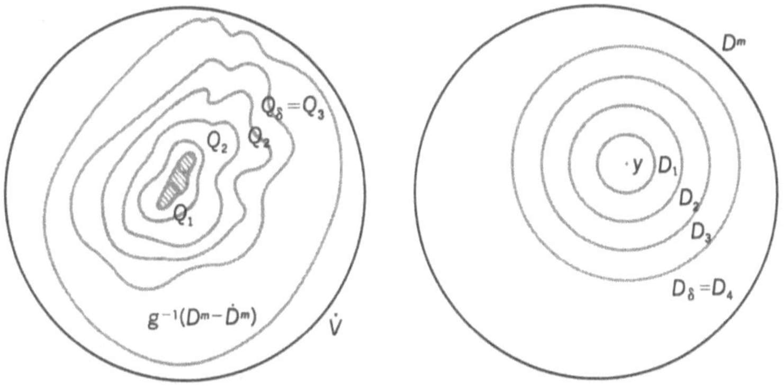

Figure 2.2

Proof. We assume $g^{-1}(D^m - \dot{D}^m) \neq \emptyset$ or there is nothing to prove. Let y be any point of $g(V) \cap (D^m - \dot{D}^m)$. Then $g^{-1}(y)$ is a compact subset of V. We can choose $\delta > 0$ so small that if D_δ is the closed disc of radius δ about y, then $D_\delta \subset D^m - \dot{D}^m$ and thus $Q_\delta = g^{-1}(D_\delta) \subset V$.

Let D_i be a closed disc about y of radius $(i/4)\delta$ for $i = 1, 2, 3, 4$, and let $Q_i = g^{-1}(D_i)$. Cover D^m by open discs of radius $\delta/16$ and choose a refinement $\mathcal{U} = \{U_\alpha\}$ of order n of the covering of $\bar{V}$ by the pre-images of these discs. Since $\bar{V}$ is compact, we assume $\mathcal{U}$ is finite and we may construct the partition of unity $P_\alpha(x) = d(x, \bar{V} - U_\alpha) / \sum_\alpha d(x, \bar{V} - U_\alpha)$ subordinated to $\mathcal{U}$. Select a point $x_\alpha \in U_\alpha$ for each α. Define a "linear approximation" to g by the formula $h(x) = \sum_\alpha p_\alpha(x) g(x_\alpha)$. Let $u \colon \bar{V} \to I$ be an Urysohn function which is 0 on $\bar{Q}_2$ and 1 on $\bar{V} - Q_3$. Blend g and h together by setting $g'(x) = u(x)g(x) + (1 - u(x))h(x)$. Observe that $g' \mid \dot{V} = g \mid \dot{V}$, that $\mid g'(x) - g(x) \mid = (1 - u(x)) \mid g(x) - h(x) \mid < \delta/8$, and that if $x \in Q_2$, then $g'(x) = h(x)$. But then the image of g' meets D_1 in a union of restrictions to D_1 of a finite number of affine subspaces of dimension $\leq n$. Thus g' omits some point of D_1. Also, since $\mid g(x) - g'(x) \mid < \delta/8$ and $g \mid (\bar{V} - Q_3) = g' \mid (\bar{V} - Q_3)$, the function $G(x,t) = tg'(x) + (1-t)g(x)$ is a homotopy of g' relative to $\dot{V}$. ∎

We now state the main theorem of this section and some of its corollaries.

Theorem 8.5. (*Cellular approximation theorem*). *Let X and Y be CW complexes, let A be a subcomplex of X, and let $f \colon X \to Y$ be a continuous map such that $f \mid A$ is cellular. Then there is a cellular map $g \colon X \to Y$ and a homotopy H of f and g such that $H(a,t) = f(a) = g(a)$ for $a \in A$.*

Before we give the proof, we state the following corollaries to this theorem.

Corollary 8.6. *Let X and Y be CW complexes, let A be a subcomplex of X, and let $f \colon X \to Y$ be a map such that $f \mid A$ is cellular. Suppose in addition there are subcomplexes $B_1, B_2, \cdots, B_n$ of X*

and $C_1, C_2, \cdots, C_n$ *of* Y *and* $f(B_i) \subset C_i$ *for each* i. *Then there is a cellular map* $g:X \to Y$ *such that* $g(B_i) \subset C_i$ *for each* i, *and* g *is homotopic to* f *relative to* A. *(This is the extension of the cellular approximation theorem to* n*-ads.)*

Corollary **8.7.** *Let* X *and* Y *be CW complexes, let* A *be a subcomplex, and let* $H:(X \times \{0\}) \cup (A \times I) \to Y$ *be a cellular map. Then there is a cellular homotopy* $H':X \times I \to Y$ *extending* H. *If in addition* $B_1, B_2, \cdots, B_n$ *and* $C_1, C_2, \cdots, C_n$ *are subcomplexes of* X *and* Y, *respectively, and* $H((B_i \times \{0\} \cup (B_i \cap A) \times I) \subset C_i$ *for each* i, *we may choose* H' *so that* $H'(B_i \times I) \subset C_i$ *for each* i.

Corollary **8.8.** *Let* X *and* Y *be CW complexes, let* A *be a subcomplex of* X, *and suppose* f, $g:X \to Y$ *are cellular maps which are homotopic via a homotopy* $H:X \times I \to Y$ *which is cellular on* $A \times I$. *Then* f *and* g *are homotopic via a cellular homotopy* H' *such that* $H' \mid (A \times I) = H \mid (A \times I)$.

Proof of theorem 8.5. We will construct g and the homotopy H of f with g inductively over the subcomplexes $A \cup X^n$. For $n=0$, we define $g^0 \mid A = f \mid A$, and for any 0-cell $\sigma \subset X - A$, define $g^0(\sigma)$ to be any 0-cell in the path component of $f(\sigma)$. Define the homotopy H^0 of $f \mid (A \cup X^0)$ with g^0 to be $H(a,t) = f(a)$ for $a \in A$, and $H^0 \mid (\sigma \times I)$ to be any path between $f(\sigma)$ and $g^0(\sigma)$ for σ a 0-cell in $X - A$.

Suppose we have constructed $g^{n-1}:A \cup X^{n-1} \to Y$ so that g^{n-1} is cellular, and we have constructed a homotopy H^{n-1} of $f \mid (A \cup X^{n-1})$ with g^{n-1} such that $H^{n-1}(a,t) = f(a) = g^{n-1}(a)$ for $a \in A$. Define

$$H':((A \cup X^{n-1}) \times I) \cup (X^n \times \{0\}) \to Y$$

by

$$H' \mid ((A \cup X^{n-1}) \times I) = H^{n-1}, \qquad H' \mid (X^n \times \{0\}) = f,$$

and use the homotopy extension theorem to extend to

$$\bar{H}:(A \cup X^n) \times I \to Y.$$

Then $\bar{H} \mid ((A \cup X^n) \times \{1\}) = g'$ is a map which extends g^{n-1} over $A \cup X^n$, but need not map n-cells into n-cells. We will deform g' on the interior of each n-cell with interior in $X - A$ into a cellular map g^n which is homotopic to g' under a homotopy relative to $A \cup X^{n-1}$. We then fit these homotopies together to obtain H^n.

Now let σ be an n-cell with interior in $X - A$, let g' denote $g' \mid \sigma$, and suppose g' maps $(\sigma, \dot{\sigma})$ into (Y, Y^{n-1}). Since $g'(\sigma)$ is compact, its carrier is a finite complex, and thus at this stage we may assume that Y is the carrier of $g'(\sigma)$ and Y is a finite complex. Let τ be any cell of top dimension in Y, where we assume that the dimension of τ is greater than that of σ. It will suffice to show that g' is homotopic relative to $\dot{\sigma}$ to a map which omits an interior point of τ; then by radial deformation from this point we will get a map g'', the image of which does not meet $\tau - \dot{\tau}$. After finitely many such deformations, we will obtain a map homotopic to g' which maps σ into Y^n.

Since $g'(\dot{\sigma}) \subset Y^{n-1}$, it follows that $(g')^{-1}(\tau - \dot{\tau}) \subset \sigma - \dot{\sigma}$. Thus if $y \in g'(\sigma) \cap (\tau - \dot{\tau})$, then $(g')^{-1}(y)$ is closed and compact in σ, is a subset of $\sigma - \dot{\sigma}$, hence is compact in $\sigma - \dot{\sigma}$. Let D^m be a neighborhood of y contained in $\tau - \dot{\tau}$ homeomorphic to a closed disc. Let $V = (g')^{-1}(D^m - \dot{D}^m)$. Then $\bar{V} \subset \sigma$, and $g' : \bar{V} \rightarrow D^m$. If $n < m$, we can apply lemma 8.4 and conclude that g' is homotopic relative to $\sigma - V$ to a map which omits a point of $\tau - \dot{\tau}$. Thus g' is homotopic relative to $\sigma - V$ to a map which does not meet $\tau - \dot{\tau}$. ∎

Proof of corollary 8.6. Select the smallest nonempty intersection of the B's, and let B_I be any such. Let B_J be any intersection of all but one of the factors of B_I and let C_J be the corresponding intersection of the C's. Apply the cellular approximation theorem to $f \mid B_J$, replacing X by B_J, A by $A \cap B_J$, and Y by C_J. Let g_J be the map so obtained. We note that the maps g_J are consistent whenever $B_J \cap B_{J'} \neq \emptyset$ because the intersection is then a minimal one. Now let B_K run over all intersections of B's with one fewer factor than the ones yielding the B_J's. For any such the homotopy extension theorem implies the existence of a map g_K' mapping B_K into C_K extending the cellular map given by g_J on $B_J \cap B_K$ for any B_J and f on $A \cap B_K$. Apply the cellular approximation theorem to get a cellular map $g_K : B_K \rightarrow C_K$ extending $f \mid A$ and $g_J \mid B_J \cap B_K$. After a finite number of such applications of the cellular approximation theorem and the homotopy extension theorem, we get a map $g':A\cup B_i\to Y$ which is cellular, extends $f\mid A$, maps B_i into C_i, and is homotopic to f. By the homotopy extension theorem, there is a homotopic g' extending f to all X, and, by the cellular approximation theorem, we may assume it to be cellular. ∎

We omit the proofs of the remaining corollaries.

Corollary **8.9.** *Let X and Y be CW complexes, let A be a subcomplex of X, and let $f:X\to Y$ be a map such that $f\mid A$ is cellular. Then there is a cellular approximation g to f and a homotopy H of f with g, both of which are carried by the minimal carrier of f.*

Proof. We proceed by induction on the dimension of cells of X which meet $X-A$. If $n=-1$, then $g^{-1}=f\mid A$ is a given cellular map and $H^{-1}:A\times I\to Y$ given by $H(a,t)=f(a)$ both belong to the minimal carrier of f.

Given $H:((X^{n-1}\cup A)\times I)\cup(X^n\times\{0\})\to Y$ which is carried by the minimal carrier E of f, we extend to $\bar{H}:(X^n\cup A)\times I\to Y$ carried by E. Note that $g'^n=H\mid((X^n\cup A)\times\{1\})$ will be carried by E. Let g^n be constructed as in the proof of 8.5.

We assert that for any cell $\sigma\subset X$, we have $H(\sigma\times I)\subset C(f(\sigma))=E(\sigma)$, and thus that $C(H(\sigma\times I))\subset E(\sigma)$. Given any point x of σ, interior to $\tau\subset\sigma$, the successive stages in the homotopy $H:g\simeq f$ move x through a sequence $f(x)=a_0, a_1, \cdots, a_k=g(x)$, such that a_i lies on the boundary of the cell containing a_{i-1} in its interior. Thus $(a_0,\cdots,a_k)$ is a c-path with points lying in cells of decreasing dimension and thus each point a_i lies in the carrier of $a_0=f(x)$. The paths joining the points traced during the homotopy also clearly lie in this carrier. Thus

$$H(\sigma\times I)\subset\bigcup\{C(f(x))\mid x\in\sigma\}\subset C(f(\sigma))=E(\sigma).\quad ∎$$

*9. ASPHERICAL CARRIER THEOREM

Let X be a CW complex and Y a space.

Definition **9.1.** If E is a carrier from X to Y such that for each cell σ of X and each k, $\pi_k(E(\sigma))=0$, then E is an *aspherical* *carrier* from X to Y. If $f:X \to Y$ is a map and if for each cell σ of X we have $f(\sigma) \subset E(\sigma)$, we say that f is carried by E. If $F:X \times I \to Y$ is a homotopy such that for each cell σ of X, $F(\sigma \times I) \subset E(\sigma)$, the homotopy F is carried by E.

THEOREM **9.2.** *Let X be a CW complex and E an aspherical carrier to Y. Let A be a subcomplex of X.*

(i) *Given any map $f:A \to Y$ carried by E, there is an extension $f':X \to Y$ carried by E.*

(ii) *Given maps $f,g:X \to Y$ and a homotopy $F:A \times I \to Y$ between $f \mid A$ and $g \mid A$ carried by E, there is an extension $F':X \times I \to Y$ of F to a homotopy of f with g which is carried by E.*

Proof. We inductively extend f over the subcomplexes $X^n \cup A$ of X. If $n = -1$, this subcomplex is A, and the corresponding map is the given f. Suppose we have extended f to $f^n:X^n \cup A \to Y$ which is carried by E. Let σ be an $(n+1)$-cell of X. If $\sigma \subset A$, define $f^{n+1} \mid \sigma = f \mid \sigma$. If σ is not a cell of A, then f^n is defined on $C(\dot{\sigma})$, and $f^n(\dot{\sigma}) \subset E(\sigma)$. Since the subspace $E(\sigma)$ has vanishing homotopy groups, if φ is the characteristic map for σ the composition $f^n\varphi:E^{n+1} \to Y$ extends to a continuous map $\psi:E^{n+1} \to Y$. The map $\psi\varphi^{-1}:\sigma \to Y$ together with f^n defines an extension of f^n over $\sigma \cup X^n \cup A$. We do this for each $(n+1)$-cell to obtain $f^{n+1}:X^{n+1} \cup A \to Y$ which extends f and is carried by E. The union of the maps f^{n+1} is an extension of f with the desired properties.

The proof of the second statement is very much like that of the first. We inductively extend the given homotopy F over the subcomplexes $(X \times \{0\}) \cup ((X^n \cup A) \times I) \cup (X \times \{1\})$. The given homotopy together with f and g defines the initial F. Assume that F has been extended to a map $F^n:(X \times \{0\}) \cup ((X^n \cup A) \times I) \cup (X \times \{1\}) \to Y$ which is carried by E. Let σ be an $(n+1)$-cell of X. If σ is a cell of A, define $F^{n+1} \mid (\sigma \times I) = F \mid (\sigma \times I)$. If σ is not a cell of A, the map F^n defines a map $(\sigma \times \dot{I}) \cup (\dot{\sigma} \times I) \to E(\sigma)$. Since $E(\sigma)$ is aspherical, this map extends over $\sigma \times I$, and we define $F^{n+1} \mid (\sigma \times I)$ to be this extension. This defines F^{n+1}, and the union of the maps F^{n+1} yields a homotopy which extends F and has the desired properties. ∎

CHAPTER III

REGULAR AND SEMISIMPLICIAL CW COMPLEXES

In this chapter we will study two special kinds of CW complexes. Regular complexes, the first topic, are interesting because they are very useful in homological calculations. As we shall see in Chapter V, they share with simplicial complexes the virtue of allowing a relatively easy calculation of the "algebraic boundary" of a cell. At the same time, they share with more general CW complexes the advantage that often a space can be represented as a regular CW complex with many fewer cells than in a simplicial decomposition. A regular CW complex can be subdivided into a simplicial complex: in this sense it is a simplicial complex in which the simplexes are more efficiently combined into closed cells. For example, the $\otimes$ product of any two regular complexes is again a regular complex. If they are simplicial complexes, then the cells of the product complex are subdivided (in a nonunique way) to give the standard product simplicial complex.

Our second interest is in certain CW complexes associated with semisimplicial complexes. Our development of this latter subject is quite limited because this is a book about CW complexes, but there is an extensive literature which motivates its study. For example, semisimplicial complexes are useful in working with the singular complex of a space (itself a semisimplicial complex) and in the investigation of the relations between the singular homology and homotopy groups of a space [11,32]. In principle if not in practice, semisimplicial theory leads to a complete solution of the problem of finding the homotopy groups of a simplicial complex: it gives the generators and relations of a presentation for each group [32]. A semisimplicial complex (an ssc) can be regarded as a generalization of an ordered simplicial complex. An element of an ssc has a dimension; one of dimension n is called an n-simplex, and like an ordered n-simplex has $n+1$ faces of dimension $n-1$ and $n+1$ degeneracies of dimension $n+1$. Generalizing from the case of an ordered simplicial complex, Milnor has associated with each ssc X a CW complex $|X|$ called its geometric realization [28]. For each nondegenerate n-simplex, x_n of X there corresponds an n-cell of $|X|$ and the faces of x_n correspond to the faces of the cell $|x_n|$. Unlike the case of a simplicial complex where each simplex injects into the realization, here we may have to identify points of Δ^n for some cells. That is, some cells of $|X|$ need not be homeomorphic to E^n.

Aside from geometrically realizing arguments in the singular complex, this construction has some other uses. Because each such geometric ssc has a subdivision into a simplicial complex and because many common topological constructions lead from simplicial complexes and maps to semisimplicial complexes, it gives a uniform method of showing that these constructions lead from polyhedra (triangulable spaces) to polyhedra. Also, the realization of the singular complex of a space gives a universal object through which maps of simplicial or semisimplicial complexes into the space can be factored. Finally, in Chapter IV we will see that the realization of the singular complex of a space enters into the study of CW homotopy type.

1. *REGULAR AND NORMAL CW COMPLEXES*

The following properties correspond to those introduced for combinatorial cell complexes in I.1.7.

Definition **1.1.** A CW complex X is *regular* if each closed cell is homeomorphic to a closed Euclidean n-cell. A CW complex is *normal* if each closed cell is a subcomplex.

We point out that if $(X,\mathcal{S})$ is regular, then we can choose a cell structure on X in which each characteristic map is a homeomorphism and for which the cells are the same subsets of X as those of the cells of $\mathcal{S}$. However, this need not be strictly equivalent in the sense of Chapter I to a cell structure defining $\mathcal{S}$.

A normal CW complex need not be regular as is shown by the decomposition of real projective space $\mathbf{RP}^n$ given in I.2.4. In this example, each cell is homeomorphic to a projective space. In the next section we will prove that a regular CW complex is a normal complex.

LEMMA **1.2.** *A CW complex X is normal if and only if for each cell σ of X, $C(\dot{\sigma})\subset\sigma$, where C is the carrier function.*

Proof. The CW complex X is normal if and only if each cell σ is a subcomplex, which is equivalent to $\sigma=C(\sigma)=(\sigma-\dot{\sigma})\cup C(\dot{\sigma})\supset C(\dot{\sigma})$ for each cell σ. ∎

LEMMA **1.3.** *A CW complex X is normal if and only if for each cell σ of X, the boundary $\dot{\sigma}$ is a subcomplex of X.*

Proof. If X is normal, both σ^n and X^{n-1} are subcomplexes of X, so that $\dot{\sigma}^n=\sigma^n\cap X^{n-1}$ is a subcomplex of X.

If for each cell σ of X, $\dot{\sigma}$ is a subcomplex, then $\sigma=(\sigma-\dot{\sigma})\cup\dot{\sigma}=(\sigma-\dot{\sigma})\cup C(\dot{\sigma})=C(\sigma)$, and σ is a subcomplex. Thus X is normal. ∎

PROPOSITION **1.4.** (i) *If A is a subcomplex of the regular (or normal) CW complex X, then A is a regular (respectively normal) CW complex.*

(ii) *If X and Y are regular (or normal) CW complexes, then $X\otimes Y$ is a regular (respectively normal) CW complex.*

The proofs are easy and are left to the reader. ∎

Definitions **1.5.** If X is a CW complex and σ,τ are cells of X, we say that τ is a *face* of σ provided $\tau\subset\sigma$. If τ is a face of σ and $\tau\neq\sigma$, we say that τ is a *proper face* of σ. A cell σ is a *principal cell* of X if it is not a proper face of any cell of X.

LEMMA 1.6. *A cell σ of a normal CW complex X is a principal cell if and only if $\sigma - \dot{\sigma}$ is open in X.*

Proof. If σ is a principal cell of X, then $X - (\sigma - \dot{\sigma}) = \mathbf{U}\{\tau \mid \tau \neq \sigma\}$, and since each of the cells τ is a subcomplex, $\mathbf{U}\{\tau \mid \tau \neq \sigma\}$ is a subcomplex of X. But then this union is closed, and $\sigma - \dot{\sigma}$ is open in X.

If $\sigma - \dot{\sigma}$ is an open subset of X, then $X - (\sigma - \dot{\sigma})$ is closed. Therefore, if $\tau - \dot{\tau} \subset X - (\sigma - \dot{\sigma})$, its closure, τ, is contained in $X - (\sigma - \dot{\sigma})$, and σ is not a proper face of any cell. Thus σ is principal. ∎

The following theorem shows that regular CW complexes are closely related to simplicial complexes.

THEOREM 1.7. *If X is a regular normal CW complex, there is a triangulation of X such that the resulting simplicial CW complex is a subdivision of X.*

Proof. We will only outline the proof, leaving it to the reader to fill in the missing details. Let $T(X)$ denote the abstract simplicial complex with vertices corresponding to the cells of X and n-simplexes members of the set $\{(\sigma_0, \sigma_1, \cdots, \sigma_n) \mid \sigma_i \text{ is a cell}, \sigma_i \neq \sigma_j$ for $i \neq j$, and $\sigma_0 \subset \dot{\sigma}_1 \subset \cdots \subset \sigma_n\}$. Let $\mid T(X) \mid$ denote the geometric realization of this complex. We define a homeomorphism $\Psi: |T(X)| \to X$ such that if σ is a cell of X, $\Psi^{-1}(\sigma)$ is a subcomplex of $\mid T(X) \mid$.

The map Ψ is constructed by first noticing that $T(X) = \mathbf{U}_n T(X^n)$ and inductively constructing $\Psi^k: \mid T(X^k) \mid \to X^k$ so that Ψ^k defines a triangulation of each cell of dimension less than $k+1$ which contains the triangulated boundary of the cell as a subcomplex. First observe than if Ψ^{n-1} is defined and σ^n is an n-cell, then $\dot{\sigma}^n$ is a subcomplex of X^{n-1}, and $(\Psi^{n-1})^{-1}(\dot{\sigma}^n)$ is a subcomplex of $\mid T(X^n) \mid$ which is homeomorphic to the $(n-1)$-sphere. The usual conical construction enables us to extend Ψ^{n-1} over $\mid T(X^n) \mid$.

Geometrically, we are performing a generalized barycentric subdivision of the complex X. The new vertices correspond to the "barycenters" of the cells of X. The new n-cells correspond to the cones over the new $(n-1)$-cells of X with vertices at the "barycenters" of the n-cells. ▮

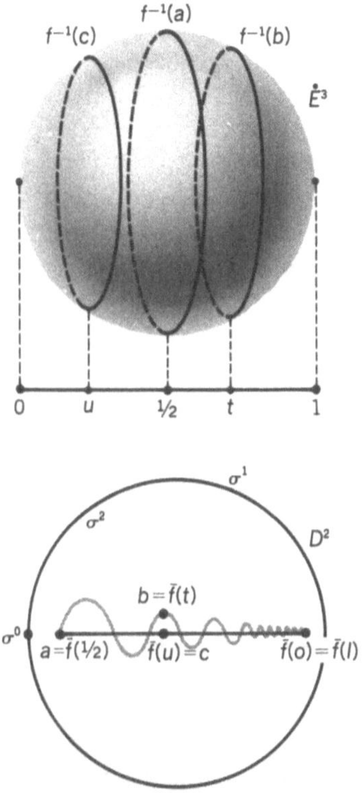

Figure 3.1

The following example shows that some restrictive condition on X is necessary in theorem 1.7.

* *Example* **1.8.** A finite 3-dimensional CW complex which cannot be triangulated.

Let $g: D^1 \to \mathbf{R}^2$ be the map $g(t) = (-t, 0)$ for $-1 \leq t \leq 0$, $g(t) = (t, t \sin(\pi/t))$ for $0 \leq t \leq 1$, which is clearly continuous; let $h: \mathbf{S}^2 \to D^1$ be the map $h(x,y,z) = x$; and let $f: \mathbf{R}^2 \to \mathbf{S}^2$ be the inverse of stereographic projection. Define $\varphi = fgh: \mathbf{S}^2 \to \mathbf{S}^2$ and $C = \varphi(\mathbf{S}^2)$.

If $X = \mathbf{S}^2 \cup_\varphi E^3$, X is compact, hence any CW decomposition of X is finite. In particular, if X can be triangulated, it can be triangulated by finitely many simplexes. Suppose X is triangulated and $\sigma^3 = A \subset X$ is the subcomplex which is the union of all 3-simplexes. If τ is a simplex and $\tau \subset \mathbf{S}^2 \cap A = \dot{\sigma}^3 = C$, then the dimension of τ is less than 2 (by an invariance of domain argument). Thus $C = \mathbf{S}^2 \cap A \subset X^1$. But the curve C has infinitely many double points, hence cannot be contained in the finite simplicial complex X^1.

* *2. REGULAR CW COMPLEXES AND INVARIANCE OF DOMAIN*

The following two theorems require the use of the theorem of invariance of domain (see 0.5), and therefore we regard them as nonelementary.

THEOREM 2.1. *If X is a regular CW complex, then X is a normal CW complex.*

Proof. By lemma 1.3, it will suffice to prove that for any n-cell σ^n of X, $\dot{\sigma}^n$ is a subcomplex. Thus we must prove that if τ^q is a cell of X and $(\tau^q-\dot{\tau}^q)\cap\dot{\sigma}^n\neq\emptyset$, then $\tau^q\subset\dot{\sigma}^n$. Since $\dot{\sigma}^n\subset X^{n-1}$, we know that if $(\tau^q-\dot{\tau}^q)\cap\dot{\sigma}^n\neq\emptyset$, then $q\leq n-1$. We prove the proposition by descending induction on q.

Let τ be an $(n-1)$-cell of X such that $(\tau-\dot{\tau})\cap\dot{\sigma}^n\neq\emptyset$. Then $(\tau-\dot{\tau})\cap\dot{\sigma}^n$ is closed in $(\tau-\dot{\tau})$. Since X is a regular CW complex, $\dot{\sigma}^n$ is homeomorphic to an $(n-1)$-sphere $\mathbf{S}^{n-1}$. Also, $\tau-\dot{\tau}$ is homeomorphic to Euclidean space $\mathbf{R}^{n-1}$. By invariance of domain the open subset $(\tau-\dot{\tau})\cap\dot{\sigma}^n$ of $\dot{\sigma}^n$ is homeomorphic to an open subset of $\tau-\dot{\tau}$. Thus $(\tau-\dot{\tau})\cap\dot{\sigma}^n$ is both open and closed in the connected space $\tau-\dot{\tau}$, and since $(\tau-\dot{\tau})\cap\dot{\sigma}^n\neq\emptyset$, we must have $\tau-\dot{\tau}\subset\dot{\sigma}^n$. Finally, since $\dot{\sigma}^n$ is closed in X, we have $\tau\subset\dot{\sigma}^n$.

Now let $C=\bigcup\{\tau\mid\tau \text{ is an } (n-1)\text{-cell of } X \text{ and } \tau\subset\dot{\sigma}^n\}\subset\dot{\sigma}^n$. The previous paragraph shows that $\dot{\sigma}^n\subset C\cup X^{n-2}$. Suppose that $\dot{\sigma}^n\subset C\cup X^q$, where $q\leq n-2$, and suppose that $\tau^q-\dot{\tau}^q\subset X^q-C$. Since $\tau^q-\dot{\tau}^q$ is open in X^q, $\tau^q-\dot{\tau}^q$ is open in X^q-C, which is open in $X^q\cup C$, so that $\tau^q-\dot{\tau}^q$ is open in $X^q\cup C$. By hypothesis, $\dot{\sigma}^n\subset C\cup X^q$, so that $(\tau^q-\dot{\tau}^q)\cap\dot{\sigma}^n$ is open in $\dot{\sigma}^n$. But $q\leq n-2$, $\tau^q-\dot{\tau}^q$ is homeomorphic to $\mathbf{R}^q$, and $\dot{\sigma}^n$ is homeomorphic to $\mathbf{S}^{n-1}$. By the corollary of 0.5 to invariance of domain, we must have $(\tau^q-\dot{\tau}^q)\cap\dot{\sigma}^n=\emptyset$. Thus $\dot{\sigma}^n\subset C\cup X^{q-1}$ and, by induction, $\dot{\sigma}^n\subset C\cup X^{-1}=C$, so that we have $\dot{\sigma}^n=C$. Also, if τ is a q-cell of X such that $(\tau-\dot{\tau})\cap\sigma^n\neq\emptyset$, then $\tau\subset C=\dot{\sigma}^n$, and $\dot{\sigma}^n$ is the space of a subcomplex. ∎

We remark that the only property of regularity used in this proof is that for each cell σ, the boundary $\dot{\sigma}$ is a topological manifold. Thus any such CW complex is a normal CW complex.

THEOREM 2.2. *Let X be a normal CW complex which is a topological n-manifold. Then*
(i) *for $q<n$, each q-cell is a face of some n-cell, and*
(ii) *if X is regular, each $(n-1)$-cell is a face of exactly two n-cells.*

Proof. (i) Suppose $q<n$ and σ^q is not a face of any n-cell of X. Choose q' as large as possible so that σ^q is a face of $\sigma^{q'}$. Now the

characteristic map φ for $\sigma^{q'}$ maps $D^{q'}-\dot{D}^{q'}$ homeomorphically onto a subset of X, and by the corollary of 0.5 to invariance of domain, we must have $q' \leq n$. Since σ^q is not a face of an n-cell, $q' < n$. But then $\sigma^{q'}$ is a principal q'-cell of X, so that $\sigma^{q'}-\dot{\sigma}^{q'}$ is open in X, and $(\varphi)^{-1}:\sigma^{q'}-\dot{\sigma}^{q'} \rightarrow D^{q'}-\dot{D}^{q'}$ is a homeomorphism. But this is impossible by the corollary to invariance of domain. We conclude that σ^q must be a face of some n-cell.

(ii) By part (i) each $(n-1)$-cell σ^{n-1} is a face of at least one n-cell. Suppose that $\sigma^{n-1} \subset \sigma_0{}^n$ and σ^{n-1} is a face of no other n-cell. Then $(\sigma^{n-1}-\dot{\sigma}^{n-1}) \cap \sigma_\lambda{}^n = \emptyset$ for $\lambda \neq 0$. Since each n-cell $\sigma_\lambda{}^n$ is the space of a subcomplex of X, $A = \bigcup_{\lambda \neq 0} \sigma_\lambda{}^n$ is the space of a subcomplex of X and is closed in X. Thus $X-A$ is open in X and $X-A = (\sigma_0{}^n - \dot{\sigma}_0{}^n) \cup B$, where B is a union of proper faces of $\sigma_0{}^n$ which contains $\sigma^{n-1}-\dot{\sigma}^{n-1}$. Let $\varphi_0:D^n \rightarrow \sigma_0{}^n$ be a characteristic homeomorphism for the cell $\sigma_0{}^n$. Since $X-A \subset \sigma_0{}^n$ and $X-A$ is open in X, by invariance of domain $(\varphi_0)^{-1}(X-A)$ is an open subset of R^n. But $(\varphi)^{-1}(X-A) = (D^n-\dot{D}^n) \cup (\varphi_0)^{-1}(B)$, and $(\varphi_0)^{-1}(B) \subset \dot{D}^n \subset D^n$ is nonempty. However, this means that $(\varphi_0)^{-1}(X-A)$ is not open and we have a contradiction. We conclude that σ^{n-1} is a face of $\sigma_\lambda{}^n$ for some $\lambda \neq 0$.

Now suppose that σ^{n-1} is a face of the n-cells $\sigma_1{}^n$ and $\sigma_2{}^n$, and $\varphi_1:D^n \rightarrow \sigma_1{}^n$ and $\varphi_2:D^n \rightarrow \sigma_2{}^n$ are characteristic homeomorphisms. Then each φ_i maps a subset of $\dot{D}^n$ homeomorphically onto $\sigma^{n-1}-\dot{\sigma}^{n-1}$ since X is regular. Define $C_i = \{x \mid x = \varphi_i(t\varphi^{-1}(p)),\ p \in \sigma^{n-1}-\dot{\sigma}^{n-1}$, and $0 < t \leq 1\}$ for $i=1,2$. Then $C_1 \cup C_2 \subset \sigma_1{}^n \cup \sigma_2{}^n$, and $C_1 \cup C_2$ is homeomorphic to $(D^{n-1}-\dot{D}^{n-1}) \times \{t \mid -1 < t < 1\}$, which is open in R^n. By invariance of domain $C_1 \cup C_2$ is an open subset of X. Thus there is a neighborhood of $(\sigma^{n-1}-\dot{\sigma}^{n-1})$ which is contained in $\sigma_1{}^n \cup \sigma_2{}^n$, and $(\sigma^{n-1}-\dot{\sigma}^{n-1}) \cap \sigma_\lambda{}^n = \emptyset$ for $\lambda \neq 1, 2$. Thus σ^{n-1} is a face of exactly two n-cells. ∎

Observe that without the assumption of regularity, this theorem certainly fails. For example, in the (normal) CW decomposition of $\mathbf{RP}^n$ given in I.2.4, the $(n-1)$-cell is a face of the unique n-cell $\mathbf{RP}^n$.

3. *SEMISIMPLICIAL COMPLEXES*

We prepare for the constructions in the succeeding sections by giving basic definitions, examples, and some simple properties.

Definition **3.1.** *A semisimplicial complex* (hereinafter referred to as an ssc) X consists of a sequence $\{X_n \mid n=0,1, \cdots\}$ of disjoint sets together with a collection of maps in each dimension n:

$d_i: X_{n+1} \to X_n$, $\quad i=0, 1, \cdots, n+1$, $\quad$ the ith *face* operator;

$s_j: X_n \to X_{n+1}$, $\quad i=0, 1, \cdots, n$, $\quad$ the jth *degeneracy* operator;

which satisfy the semisimplicial identities:

(1) $d_i d_j = d_{j-1} d_i$, $\quad i<j$;

(2) $d_i s_j = s_{j-1} d_i$, $\quad i<j$;

(3) $d_i s_j = 1$ (identity), $\quad i=j, j+1$;

(4) $d_i s_j = s_j d_{i-1}$, $\quad i>j+1$;

(5) $s_i s_j = s_{j+1} s_i$, $\quad i \leq j$.

The elements of X_n are called the *n-simplexes* of X.

If X and Y are ssc's, a *semisimplicial map* (an ss map) is a map $f: X \to Y$ mapping X_n into Y_n for each n and such that $d_i f(x) = f(d_i x)$, $s_j f(x) = f(s_j x)$ for each simplex x of X and each map d_i, s_j defined on x. A simplex of form $s_i x$ is *degenerate*; one not of this form is *nondegenerate.*

Example **3.2.** If X is a topological space, let $S_n(X)$ be the collection of all continuous maps of the standard n-simplex Δ^n into X, and let $S(X)$ be the sequence of the $S_n(X)$. This is the *singular complex* of X. Define $d_i^*: \Delta^n \to \Delta^{n+1}$ to be the map sending $(t_0, \cdots, t_n)$ to $(t_0, \cdots, t_{i-1}, 0, t_i, \cdots, t_n)$ and $s_j^*: \Delta^n \to \Delta^{n-1}$ the one sending $(t_0, \cdots, t_n)$ to $(t_0, \cdots, t_{j-1}, t_j + t_{j+1}, t_{j+2}, \cdots, t_n)$. Let $d_i: S_n(X) \to S_{n-1}(X)$ be the map sending the n-simplex $x_n: \Delta^n \to X$ into $x_n d_i^*$, and $s_j: S_n(X) \to S_{n+1}(X)$ to be the map sending x_n into $x_n s_j^*$. Then $S(X)$ is an ssc. If $f: X \to Y$ is a continuous map, then the map f induces an ss map $S(f): S(X) \to S(Y)$ by sending x_n into $f x_n$. Obviously S is a functor.

Example **3.3.** Any subset of the singular complex of a space which is closed under all permissible face and degeneracy operators is an ssc. Such a subset of an ssc is called a *subcomplex*. If A is a subspace of X, then $S(A)$ is a subcomplex of $S(X)$. (In fact, any ssc is a subcomplex of the singular complex of some space.)

Example **3.4.** Let K be a simplicial complex with vertex set $V = \{v_\alpha\}$. Let # denote some partial order on V which has the property that any set of vertices which lies in a simplex of K is linearly ordered by #. Let $K_n^{\#}$ denote the collection of all nondecreasing linearly ordered sequences $\langle v_0, \cdots, v_n\rangle$ of elements of V which lie in a simplex of K, and let

$$d_i\langle v_0, \cdots, v_n\rangle = \langle v_0, \cdots, v_{i-1}, v_{i+1}, \cdots, v_n\rangle,$$

$$s_j\langle v_0, \cdots, v_n\rangle = \langle v_0, \cdots, v_j, v_j, \cdots, v_n\rangle.$$

Then the sequence $K^{\#}$ of the $K_n^{\#}$ together with these maps d_i and s_j is an ssc. Given $K^{\#}$ we can clearly recover the simplicial complex K. We will refer to such an ssc as an ordered simplicial complex, and in fact, unless the order # is absolutely essential, we will refer to the ssc as K. If $f:K\to L$ is a simplicial map, the obvious candidate for an ss map of the associated ssc's, the map f' that sends $\langle v_0, \cdots, v_n\rangle$ into $\langle f(v_0), \cdots, f(v_n)\rangle$, will be an ss map if and only if f is nondecreasing with respect to the two vertex orderings. Given a simplicial map, it is not hard to see that such a pair of orderings can always be found.

Example **3.5.** The *standard ss n-simplex* $\sum(n)$ is the ordered simplicial complex whose vertices are the elements $\{0,1, \cdots, n\}$ with the usual ordering, and whose k-simplexes are all nondecreasing sequences with $k+1$ elements. The ith face operator is given by deleting the $(i+1)$-st vertex, and the jth degeneracy by repeating the $(j+1)$-st vertex.

Note that the ss maps $\varphi: \sum(n) \to \sum(m)$ correspond to monotone functions from $\{0,1, \cdots, n\}$ to $\{0,1, \cdots, m\}$. In particular we define:

$$d_j^*: \textstyle\sum(n) \to \sum(n+1)$$

$$s_j^*: \textstyle\sum(n) \to \sum(n-1)$$

by the following vertex maps:

$$d_j^*(k) = k, \quad \text{if} \quad k < j;$$

$$= k+1, \quad \text{if} \quad k \geq j;$$

$$s_j^*(k) = k, \quad \text{if} \quad k \leq j;$$

$$= k-1, \quad \text{if} \quad k > j.$$

We call d_i^* the ith face map and s_j^* the jth degeneracy map.

We use the same notation as in the case of $d_i^*: \Delta^n \to \Delta^{n+1}$ and s_j^*, the maps of geometric simplexes. In any context it should be clear which one we mean.

If X is an ssc and x an n-simplex, there is an ss map $\varphi_x: \sum(n) \to X$ defined by $\varphi_x(\sigma) = x$, where σ is the fundamental n-simplex $\{0,1, \cdots, n\}$. By analogy, we call this the characteristic map of the n-simplex x. If we take one copy of $\sum(n)$ for each n-simplex x of X, then the disjoint union $M'(X)$ of all these simplexes is an ssc and the characteristic maps of the simplexes of X determines an ss surjection of M' onto X.

Example **3.6.** Let K be any abstract simplicial complex. Associated with it is the partially ordered simplicial complex K', its *barycentric subdivision*, defined as follows. We partially order the simplexes of K by $\sigma < \tau$ if τ is a face (= subset) of σ. The vertices of K' are the simplexes of K with this partial order. A linearly ordered nondecreasing set of $k+1$ vertices of K' is a k-simplex of K'. If $f: K \to L$ is a simplicial map, then $f': K' \to L'$ is the order preserving map which sends the vertex of K' associated with the simplex σ of K into the vertex of L' associated with the simplex $f(\sigma)$ of L. Subdivision is a functor from simplicial complexes to ordered simplicial complexes of the type discussed in 3.4. We remark that there is a standard linear homeomorphism of $|K'|$ with $|K|$ (see, for example, [10, 35]), but that this homeomorphism is not natural with respect to all simplicial maps. If f is an injection map, then $|f|$ is identified with $|f'|$, but not otherwise. A simple example is the case of a map of Δ^2 into Δ^1 by identifying the second and first vertices. The centroid of Δ^2 is mapped into the midpoint of Δ^1 by $|f'|$ and into a trisecting point by $|f|$.

Under suitable conditions on the dimension n and the indices of the operators involved, we can iterate the semisimplicial operations on an n-simplex. We will denote $d_i \cdots d_j$ by d_I, where $I = (i, \cdots, j)$ is the ordered set of indices, and call such an iteration or the identity map ($I = \emptyset$) a face operator; similarly $s_i \cdots s_j$ will be denoted by s_I and be called a degeneracy operator. (Again this includes the case $I = \emptyset$, s_I then being the identity.)

Because of the ss identities, we have to distinguish between two ss operators, iterations of face or degeneracy maps, being equal and being identical, and must consider the problem of how to show that nonidentical operators are or are not equal. This is solved as follows. Given an ss operator ψ, by application of the ss identities, ψ can be expressed as a product $D \cdot F = s_J d_I$, where $J = (j_1, \cdots, j_p)$, $I = (i_1, \cdots, i_q)$, and $j_1 > \cdots > j_p$ and $i_1 < \cdots < i_q$, the *canonical form* of ψ. Moreover, it can be proved that this word is unique—two canonical words in the ss operations of this form which can be transformed into each other by application of the ss identities are identical. For example, one can give an algorithm for reduction of an ss word to this form. Pick the face operator farthest to the left which is out of place. Apply the ss identities to move it to the right until it cancels a degeneracy or it reaches a point where to the right there are only face operators with larger indices. Then pass to the next face operator to the right and repeat. After moving all the face operators into proper position, a similar rule aligns the degeneracy operators. This algorithm has the two properties: if ψ is in canonical form to begin with, it yields ψ; and each time a word is transformed by application of the ss identities, the algorithm yields the same canonical word (one checks each of the five cases). Thus if two words can be transformed into each other, at each stage the transforms give the same canonical word; if the words are canonical to begin with, they must therefore be identical.

Finally, each degenerate simplex can be expressed uniquely as

Dx, where D is some degeneracy operator and x is nondegenerate. This follows from the fact (another exercise in the ss identities) that if D and D' are unequal degeneracy operators in canonical form, then there is a face operator F such that $FD=1$ and $FD'=D''F''$, where $D''\neq 1$. Thus, if $Dx=D'x'$ are two different expressions for a degenerate simplex, with x,x' nondegenerate, then $x=FD'x'=D''(F''x')$, a degenerate simplex, which is impossible.

4. THE REALIZATION OF SEMISIMPLICIAL COMPLEXES

Given an ssc X, we will construct a normal CW complex $|X|$ called the *geometric realization* of X. It will have one n-cell for each nondegenerate n-simplex of X, and the faces of this cell will correspond to the nondegenerate faces of the given simplex. Given an ss map $f:X\to Y$, we will associate a continuous regular cellular map of the corresponding realizations which will map a cell corresponding to the nondegenerate simplex x onto the cell associated with $f(x)$ if this simplex is nondegenerate, or if not, onto the cell associated with the nondegenerate simplex of which $f(x)$ is a degeneracy.

To each n-simplex x_n of the ssc X, associate a copy (Δ^n,x_n) of the standard n-simplex. If $t=(t_0,\cdots,t_n)$ is a point of Δ^n, the corresponding point of (Δ^n,x_n) will be denoted by (t,x_n). Let $M(X)$ be the topologized disjoint union of all the copies (Δ^n,x_n). We generate an equivalence $\mathcal{R}$ on $M(X)$ by defining elementary equivalences:

$$(d_i{}^*t,x_n)\sim(t,d_ix_n), \quad \text{for all} \quad t\in\Delta^{n-1}$$

$$(s_j{}^*t,x_n)\sim(t,s_jx_n), \quad \text{for all} \quad t\in\Delta^{n+1}$$

and then defining $(t,x)\sim(u,y)$ if there is a finite chain of such elementary equivalences beginning at (t,x) and ending at (u,y). Let $|X|$ be the quotient space of $M(X)$ by $\mathcal{R}$, let $\eta:M(X)\to|X|$ be the quotient map, and denote the image of (t,x) by $[t,x]=\eta(t,x)$.

We make two preliminary technical observations. If $\psi=s_Jd_I$ is an ss operation that maps X_p to X_q, then $\psi^*=d_{\bar{I}}{}^*s_{\bar{J}}{}^*$ where $\bar{I}$

and $\bar{J}$ are the sequences I and J reversed, since for any singular simplex x, ψ and ψ^* are related by $\psi x = x\psi^*$ (composition). Moreover, ψ^* is a simplicial map sending Δ^q to Δ^p.

Secondly, if t is any point of Δ^n, then t has a unique expression F^*t', where F is a face operator and t' is an interior point of its simplex Δ^q. This expression consists of the unique open simplex or vertex of Δ^n in which t lies together with its coordinates in that simplex. The point t' can be found from $t = (t_0, \cdots, t_n)$ by deleting all zero coordinates; this gives $t' = (t_0', \cdots, t_q')$, an interior point of Δ^q. The map F^* is the one from Δ^q to Δ^n that replaces these zeros (see 3.2).

LEMMA **4.1.** *In each* $\mathcal{R}$ *equivalence class there is one and only one element* (t,x_n), *where* t *is an interior point of* Δ^n *and* x_n *is nondegenerate. Call such a point an* irreducible *representative of its class. If* (t,x) *is irreducible, then any other* (u,x), *where* u *is interior to* Δ^n, *is also.*

Proof. We define a (noncontinuous) selection function θ which associates with any point of $M(X)$ an equivalent irreducible point. Given the point (t,x), first express $t = F^*t'$, where F is a face operator and t' is an interior point. With this F, express $Fx = Dx'$, where D is either a degeneracy operator or the identity and x' is nondegenerate. Then $\theta(t,x) = (D^*t',x')$. Because t' is an interior point, its image under a simplicial map D^* is also an interior point, and therefore $\theta(t,x)$ is an irreducible point of $M(X)$.

To prove the uniqueness of this representation, we will prove that if (t,x) and (t',x') differ by a single elementary equivalence, then $\theta(t,x) = \theta(t',x')$. If both (t,x) and (t',x') are irreducible, they are equal to their θ values; if they are equivalent, they must then be identical. Suppose first the two points are (d_i^*t,x) and (t,d_ix). Express $d_i^*t = F^*u$; the corresponding expression for the other point is $t = F_1^*u$, where $F = F_1d_i$. For the expression $d_i^*t = F^*u$ is unique when u is an interior point, and thus from the fact that $t = F_1^*u$, we derive the fact that $d_i^*t = d_i^*F_1u_1$, so that $F^* = d_i^*F_1^*$, and thus $F_1d_i = F$ and $u = u_1$. Hence the first step in reduction to $\theta(t,x)$ and $\theta(t',x')$ leads to (u,Fd_ix) and (u,F_1x), respectively. Since these are equal, the values of the points under θ are also. The other type of elementary equivalence is handled in the same way.

If (t,x) is irreducible, then x is nondegenerate, and therefore each point (u,x) with u an interior point of its simplex is irreducible. ∎

We introduce some notation that we will use through lemma 4.5. If $\sigma = \sigma^p = F^*\Delta^p$ is a p-dimensional face of Δ^n, and $x = x_n$ is an n-simplex, we denote the corresponding cell of (Δ^n,x_n) by (σ^p,x_n) or (σ,x) and similarly its interior by $(\sigma^p - \dot\sigma^p,x_n)$ or $(\sigma - \dot\sigma,x)$.

Corollary 4.2. *Let F, σ^p, x_n be as above. If t is an interior point of Δ^p and (F^*t,x_n) has irreducible representative $(\bar t,\bar x_q)$, then $\eta(\Delta^q - \dot\Delta^q,\bar x_q) = \eta(\sigma^p - \dot\sigma^p,x_n)$ and $\eta(\sigma^p,x_n) = \eta(\Delta^q,\bar x_q)$.*

Proof. By the definition of θ, $Fx_n = D\bar x_q$ for some degeneracy operator D, and $D^*t = \bar t$. Then for any other u in Δ^p, (F^*u,x_n) is equivalent to $(D^*u,\bar x_q)$, and thus $\eta(\sigma^p,x_n) = \eta(\Delta^q,\bar x_q)$. Since D^* is a simplicial map, it sends interior points onto interior points, and therefore $\eta(\sigma^p - \dot\sigma^p,x_n) = \eta(\Delta^q - \dot\Delta^q,\bar x_q)$. ∎

We note that $M(X)$ has a normal CW structure, its cells being the union of the cells of the simplexes (Δ^n,x_n).

Lemma 4.3. *The equivalence relation $\mathcal{R}$ is cellular.*

Proof. Let (σ^p,x_n) be an arbitrary cell of $M(X)$, and let F, σ^p, x_n, $\bar x_q$ be as above. By corollary 4.2 the saturation of $(\sigma^p - \dot\sigma^p,x_n)$ is equal to the saturation of its irreducible cell $(\Delta^q - \dot\Delta^q,\bar x_q)$, and a second application of the corollary shows that if any open cell $(\tau^r - \dot\tau^r,x')$ meets $\eta^{-1}\eta(\Delta^q - \dot\Delta^q,x_q)$, then it is wholly contained in it. Thus the saturation of an open cell of $M(X)$ is a union of open cells.

If (σ^p,x_n) is an $\mathcal{R}$-minimal p-cell, then its equivalent irreducible cell must be of the same dimension and thus be $(\Delta^p,\bar x_p)$, so that

$\bar{x}_p = Fx_n$ by corollary 4.2. Thus if $(F^*\Delta^p, x_n) = (\sigma, x)$ is any minimal cell, its characteristic map $\varphi_{(\sigma,x)} = \varphi$ is related to the characteristic map $\varphi_{\bar{x}}$ of its equivalent irreducible cell $(\Delta^p, Fx_n) = (\Delta^p, \bar{x}_p)$ by $\varphi_{(\sigma,x)} = (\varphi_{\bar{x}_p})F^*$. Since $F^*: (\Delta^p, \bar{x}_p) \to (\sigma^p, x_n)$ is a bijection, this is condition (3) of definition I.6.1. Also, this shows that $\varphi(\Delta^p) = \varphi_{\bar{x}}(\Delta^p)$ and, since $\varphi \mid (\Delta^p - \dot{\Delta}^p, \bar{x}_p)$ is bijective, that each $\varphi \mid (\sigma^p - \dot{\sigma}^p, x_n)$ is bijective, for each minimal cell (σ^p, x_n). This is condition (ii) of definition I.6.1. ∎

Corollary **4.4.** *If x_n is a nondegenerate n-simplex of X, then (Δ^n, x_n) is $\mathcal{R}$-minimal in $M(X)$. If (σ^n, x_p) is $\mathcal{R}$-minimal in $M(X)$, then it is equivalent to a unique irreducible simplex (Δ^n, x_n).* ∎

Lemma **4.5.** *If $\sigma^p = \sigma$ is any face of any simplex of $M(X)$, and if $(\Delta^q, \bar{x}_q)$ is its unique irreducible (and $\mathcal{R}$-minimal) $\mathcal{R}$-equivalent representative, then the map $\varphi_\sigma: \sigma \to |X|$ factors into*

$$\sigma \xrightarrow{\psi} (\Delta^q, \bar{x}_q) \xrightarrow{\varphi_{\bar{x}}} |X|$$

for ψ a linear (and therefore continuous) map of simplexes.

Proof. Let σ be the set of all points (F^*t, x_n), where F is a face operator and $F^*: \Delta^p \to \Delta^n$ is the corresponding linear map embedding Δ^p as a face of Δ^n. The interior points of σ are such points with t interior to Δ^p. If θ is the selection function of lemma 4.1, suppose for an interior point (F^*t, x_n) of σ that $\theta(F^*t, x_n) = (u, \bar{x}_q)$, where $Fx_n = D\bar{x}_q$, D is a degeneracy operator, $\bar{x}_q$ is irreducible, and $D^*t = u$. Then the map $\psi: (F^*t, x_n) \to (D^*t, \bar{x}_q)$ is a simplicial map sending all σ onto $(\Delta^q, \bar{x}_q)$ and clearly $\varphi_{\bar{x}} \cdot \psi = \varphi_\sigma$. ∎

Theorem **4.6.** *The geometric realization $|X|$ is a normal CW complex with one n-cell corresponding to each nondegenerate n-simplex x_n of X. The characteristic map of the cell $|x_n|$ corresponding to x_n is the composition of the inclusion of (Δ^n, x_n) into $M(X)$ followed by the projection onto the quotient space $|X|$.*

Proof. Corollary 4.4 and lemma 4.5 show that the conditions of II.5.6 are satisfied. ∎

Let $f:X \to Y$ be an ss map. Then f induces a map $M(f)$ sending $M(X)$ into $M(Y)$ by $M(f)(t,x_n) = (t,f(x_n))$. Since f is an ss map, it commutes with the operators d_i and s_j and therefore carries points equivalent in $M(X)$ into points equivalent in $M(Y)$. It therefore determines a map of the quotient sets $|X|$ into $|Y|$. The *geometric realization* $|f| : |X| \to |Y|$ is the map induced by $M(f)$.

Theorem 4.7. *The geometric realization of an ss map is a continuous regular cellular map.*

Proof. Clearly $M(f)$ is a continuous map for it maps a copy (Δ^n,x) of Δ^n into another $(\Delta^n,f(x))$ by the identity map and therefore is continuous on each simplex of $M(X)$, and thus on all of $M(X)$. A map of quotient spaces induced by a continuous map of the overlying spaces is continuous (in the quotient topology). Obviously $|f|$ is regular and cellular. ∎

Theorem 4.8. *The operation of assigning its realization to an object or map of the category $\mathcal{C}$ of ssc's and maps is a functor from $\mathcal{C}$ to the category of topological spaces and continuous maps.* ∎

Lemma 4.9. *Let K be a simplicial complex (with vertex order #). Then there is a regular cellular homeomorphism η_K of $|K^\#|$ onto $|K|$, the geometric realization of the simplicial complex K with the CW topology. If $f:K \to L$ is a simplicial map, then (for any compatible vertex orders) the map $|f^\#|$ is the geometric realization of the simplicial map f: i.e., the following diagram is commutative:*

$$\begin{array}{ccc} |K^\#| & \xrightarrow{|f^\#|} & |L^\#| \\ \downarrow{\scriptstyle \eta_K} & & \downarrow{\scriptstyle \eta_L} \\ |K| & \xrightarrow{|f|} & |L|. \end{array}$$

Hence the realization functor is essentially an extension of the simplicial realization functor.

Proof. Let $\{v_0, \cdots, v_n\}$ be a simplex of K, and let $t_0v_0+\cdots+t_nv_n$ be the point of the simplex of $|K|$ which has vertices v_i and corresponding barycentric coordinate t_i. We also interpret this expression in case not all the vertices are distinct by adding together the t_i which appear as coefficients of any particular vertex. Thus, in every case, whether a simplicial map $f:K\to L$ does or does not identify vertices of this simplex, we always have

$$f(t_0v_0+\cdots+t_nv_n)=t_0f(v_0)+\cdots+t_nf(v_n).$$

If $\langle v_0, \cdots, v_n\rangle$ is any n-simplex of $K^{\#}$, let η_K map the point $[(t_0, \cdots, t_n), \langle v_0, \cdots, v_n\rangle]$ into the point $t_0v_0+\cdots+t_nv_n$. This map is well defined, for

$$\begin{aligned}\eta_K((t_0, \cdots, t_n), s_j\langle v_0, \cdots, v_{n-1}\rangle)\\ &=\eta_K(t_0v_0+\cdots+(t_i+t_{i+1})v_i+\cdots+t_nv_{n-1}), \langle v_0\cdots v_{n-1}\rangle)\\ &=\eta_K(s_j{}^*(t_0, \cdots, t_n), \langle v_0, \cdots, v_{n-1}\rangle),\end{aligned}$$

and, similarly, the value of η_K is unchanged by the other type of elementary equivalence. Therefore it does not depend on the choice of representative and is well defined. It is clearly continuous and bijective, and, since this is a map between CW complexes, it is a homeomorphism.

Given such a map f,

$$\begin{aligned}\eta_L\,|f^*|\,((t_0, \cdots, t_n), \langle v_0, \cdots, v_n\rangle)\\ &=\eta_K((t_0f(v_0)+\cdots+t_nf(v_n)), \langle fv_0, \cdots, fv_n\rangle)\\ &=|f|\,\eta_K((t_0, \cdots, t_n), \langle v_0, \cdots, v_n\rangle). \quad \blacksquare\end{aligned}$$

LEMMA **4.10.** *The subcomplexes of the ssc X are in one-to-one correspondence with the subcomplexes of the CW complex $|X|$.*

Proof. To each subcomplex A of X corresponds the sub-CW-complex $|A|$ of $|X|$. Given a sub-CW-complex B of $|X|$, let B' be the nondegenerate simplexes of X corresponding to the cells of B; this set contains all the nondegenerate faces of any of its elements. Let $\bar{S}(B)$ be the collection of all Dx, where D is either the identity or a degeneracy map, and x is an element of B'. Then $\bar{S}(B)$ is a subcomplex of X (since it is closed under face and degeneracy operations). Clearly $\bar{S}(|A|) = A$ and $|\bar{S}(B)| = B$, since two subcomplexes of $|X|$ and X with the same cells and nondegenerate simplexes, respectively, are identical. ▮

PROPOSITION **4.11.** *Let X be a space. Define $j_X\colon |S(X)| \to X$ by $j_X([t,x]) = x(t)$. Then j_X is continuous. It is natural in the sense that the following diagram is commutative:*

$$\begin{array}{ccc} |S(X)| & \xrightarrow{|S(f)|} & |S(Y)| \\ \downarrow{\scriptstyle j_X} & & \downarrow{\scriptstyle j_Y} \\ X & \xrightarrow{f} & Y. \end{array}$$

(That is, if we let R be the realization functor and S the singular complex functor, then j is a natural transformation of the composite functor RS with the identity.)

Proof. That the value of j_X is independent of the choice of coordinates (t,x_n) in $M(X)$ follows by an argument similar to the argument that θ is well defined in 4.1. The continuity of j_X then follows from the facts that $j_X\eta\colon M(S(X)) \to X$ is continuous and that $\eta\colon M(S(X)) \to |S(X)|$ is a quotient map. ▮

PROPOSITION **4.12.** *Let K be any ssc and let $g\colon |K| \to X$ be any map. Then there is a unique ss map $\bar{g}\colon K \to S(X)$ such that $g = j_X|\bar{g}|$. Moreover, this universal condition uniquely determines $|S(X)|$ and j_X from among all maps $p\colon |X'| \to X$ of ssc's into X up to equivalence.*

Proof. Given g, we define $\bar{g}$ as follows. Let $|\sigma|$ be any n-cell of

$|K|$ and $\varphi_\sigma: \Delta^n \to |\sigma|$ its characteristic map, the restriction of η to (Δ^n, σ) in $M(K)$. If $t \in |\sigma|$, let $\bar{g}(t) = \eta(\varphi_\sigma^{-1}(t), g\varphi_\sigma) \in |S(X)|$. It is easy to check that $\bar{g}: K \to S(X)$ is well defined, and it is clear from the definition that $g = j_X |\bar{g}|$.

To see that this map is unique, suppose that g' were a second ss map from K to $S(X)$ such that $g = j_X |g'|$. Then for any $|\sigma|$ and any point t of $|\sigma|$, $\bar{g}(\sigma)(t) = g(t) = j_X[\varphi_\sigma^{-1}(t), g'(|\sigma|)] = g'(|\sigma|)(t)$. Thus $\bar{g}(\sigma)$ and $g'(\sigma)$ are the same singular n-simplex.

If X' were any other ssc and $j': |X'| \to X$ a continuous map with the universal property above, then there is a homeomorphism $h: |S(X)| \to |X'|$ such that $j_X = j'h$. In fact, h is the realization of an ss isomorphism k. To construct h (or k), let x be any non-degenerate n-simplex of $S(X)$, and consider $|x| : \Delta^n \to X$. By the universal condition there is a unique $x': \sum(n) \to X'$ such that $j'x' = x$. Let $k(x) = x'$. It is easy to check that k defines a map of $S(X)$ into X' such that $j_X |k| = j'$. Similarly we can construct $k': X' \to S(X)$ such that $j' = j_X |k|$. It follows that if $x: \Delta^n \to X$ is any singular simplex, then $j_X |k'kx| = j' |kx| = j_X |x|$. Therefore, by the uniqueness of x', $k'kx = x$, each x, and therefore $k'k =$ identity. Similarly, $kk' =$ identity. ∎

Applying this to the case where g is the identity map $i: |X| \to |X|$, we see that any ssc X is identified to a subset $\bar{\imath}(X)$ of $S(|X|)$ in such a way that $j_X |\bar{\imath}| =$ identity. This embedding is easily shown to be natural with respect to ss maps: if $|g| : |X| \to |Y|$, then $\bar{\imath}_Y g = S(|g|)\bar{\imath}_X$.

5. *SEMISIMPLICIAL CONSTRUCTIONS*

We will examine three topological constructions that correspond to ss constructions.

(1) Products. If X, Y are ss complexes, their product is the ssc whose set of n-simplexes $(X \times Y)_n$ is $X_n \times Y_n$, and whose face and degeneracy maps are given by operation on the coordinates: $d_i(x \times y) = d_i x \times d_i y$; $s_j(x \times y) = s_j x \times s_j y$. The set theoretic projection maps p_1, p_2 are thus *ss* maps. If X and Y are singular complexes of spaces, for example, the singular complex of their product is $X \times Y$, and the projection maps induce p_1 and p_2 (for the maps of Δ^n into the product of two spaces are in one-to-one correspondence with the pairs of maps into the two factors, etc.).

Let X,Y be ssc's. Let $|p_1| : |X \times Y| \to |X|$, $|p_2| : |X \times Y| \to |Y|$ induce the map $\alpha_{XY}: |X \times Y| \to |X| \times |Y|$ by mapping a point w into $|p_1|(w) \times |p_2|(w)$.

Lemma 5.1. *The map* α_{XY} *is a bijection.*

Proof. We will construct an inverse to $\alpha = \alpha_{XY}$. Let $[t,x_n] \times [r,y_m]$ be a point of $|X| \times |Y|$, say $t = (t_0, \cdots, t_n)$, $r = (r_0, \cdots, r_m)$. We will present an algorithm for finding a point $[(u_0, \cdots, u_{n+m}), Dx \times D'y]$ for which $D^*(u_0, \cdots, u_{n+m}) = (t_0, \cdots, t_n)$ and $D'^*(u_0, \cdots, u_{n+m}) = (r_0, \cdots, r_m)$. This point will be mapped by α onto the given point.

Let $N_0, \cdots, N_{n+m}$ be the set of numbers

$$\{t_0, t_0 + t_1, \cdots, \textstyle\sum_{i<n} t_i, r_0, \cdots, \textstyle\sum_{i<m} r_i, 1\}$$

arranged in order of size. We assert that there are degeneracy operators D,D' such that the result of applying D^*,D'^*, respectively, to the $(n+m+1)$-tuple $(N_0, N_1 - N_0, \cdots, N_{n+m} - N_{n+m-1})$ is $(t_0, \cdots, t_n)$ and $(r_0, \cdots, r_m)$. To find D^*, for example, suppose $\sum_{j\leq i} t_j$ is the kth and $\sum_{j\leq i+1} (t_j)$ the k'th of the differences. By applying the operation $s_k^* \cdots s_{k'-1}^*$ we will get a sequence of numbers in which the kth place will be occupied by $N_{k'} - N_k = t_{i+1}$. By applying similar operators to the sequence $\{N_j\}$ we can thus recover successively each of the t_i. Let D^* be the composition of all these operators. Similarly, let D'^* be the degeneracy operator which recovers the sequence $\{r_j\}$. Let $(u_0, \cdots, u_{n+m}) = (N_0, \cdots, N_{n+m})$. This constructs a point $[(u_0, \cdots, u_{n+m}), Dx \times D'y]$ which α maps onto the given point.

One can check (an exercise) that this point is independent of the representation chosen for the given point and that it defines a two-sided inverse to α. ∎

COROLLARY **5.2.** *The map* α *is continuous and bijective. It is a homeomorphism if and only if* $|X| \times |Y|$ *is a CW complex* (see II.5.1). ∎

Let α' be the map $\alpha: |X \times Y| \to |X| \otimes |Y|$ obtained by giving $|X| \times |Y|$ the CW topology.

COROLLARY **5.3.** *The map* α' *is a natural homeomorphism.*

Proof. We have only to check that α' is continuous when the product topology is refined to the CW topology. But α' maps the cell $|x \times y|$ into the product $|x| \times |y|$; this subset has the same relative topology in either case. Since $\alpha' | (|x \times y|) = \alpha | (|x \times y|)$, α' is continuous on each cell, and therefore is continuous on all of $|X \times Y|$. ∎

We remark that if K,L are simplicial complexes, and $\#, \#'$ any suitable vertex orderings, then

$$|K^{\#} \times L^{\#'}| \xrightarrow{\alpha} |K| \times |L|$$

is a standard triangulation of the product of the simplicial complexes [10]. It depends, of course, on which orderings are chosen.

We also remark that the diagonal map $|X| \to |X \times X|$ is a cellular map, in fact the realization of an ss map.

(2) **Quotients.** If (X,A) is an ss pair, then the quotient complex X/A is the one with n-simplexes $(X/A)_n$ the quotient of the set X_n by the set A_n (identifying all the elements of the subset A_n to a single element $[A_n]$). Let the image of an element x of X be denoted by $[x]$. The face and degeneracy operators are the quotient operations: $d_i[x] = [d_i x]$, $s_j[x] = [s_j x]$. Let $p: X \to X/A$ be the quotient map; it is an ss map.

PROPOSITION **5.4.** *The map* $|p| : |X| \to |X/A|$ *induces a homeomorphism* $\alpha: |X/A| \to |X|/|A|$ *such that* $\alpha |p| = \bar{p}$, *where* p *is the quotient map of* $|X|$ *onto* $|X|/|A|$.

Proof. Since p is a surjection, $|p|$ is also. If a set C of $|X/A|$ has the property that $|f|^{-1}(C) \cap |x|$ is closed and thus compact for each $x \in X$, then $|f|(|f|^{-1}(C) \cap |x|) = C \cap |f(x)|$ is closed for each $f(x) \in Y$. Thus if $f: X \to Y$ is surjective, $|f|$ is a quotient map. We have only to check that $|p|$ identifies two points if and only if p does.

Let $[t, x_n]$ be the irreducible representation of a point outside $|A|$. Then $x_n \notin A$, and $p(x_n)$ is nondegenerate. Thus if $[t, x_n]$ and $[u, y_m]$ are irreducible representations of points with the same $|p|$ image and $[t, x_n]$ lies outside $|A|$, then $[t, p(x_n)] = [u, p(y_m)]$ and $[t, p(x_n)]$ is the irreducible representative of $[u, p(y_m)]$. Thus $p(y_m) = p(x_n)$ so that $y_m \notin A$, and thus $[u, p(y_m)]$ is also irreducible. Hence $u = t$ and $p(x_n) = p(y_m)$; since x_n and y_m lie outside A, this means that $x_n = y_m$, and $[t, x_n] = [u, y_m]$.

Thus $|p|$ identifies two points if and only if they both lie in $|A|$, so that $|p|$ and $\bar{p}$ define the same quotient space. ▮

Corollary **5.5.** *The quotient of a simplicial complex by a subcomplex is (the geometric realization of) an ssc.* ▮

Let G be a group of ss isomorphisms of the ssc X. Each $g \in G$ has a realization $|g|: |X| \to |X|$ to a regular cellular homeomorphism of $|X|$. If no simplex of X is mapped into itself by g, then it is easy to check that $|g|$ has no fixed points.

We define the quotient X/G to be the ssc whose k-simplexes are the equivalence classes under G-operation of k-simplexes of S, with *ss* operations as follows. If $[x]$ is the class of the simplex x, then $d_i[x] = [d_i x]$, and $s_j[x] = [s_j x]$. The quotient map $p: X \to X/G$ is an ss map.

Proposition **5.6.** *There is a regular cellular homeomorphism* $h: |X|/G \to |X/G|$ *such that* $|p| = hp'$, *where* $p': |X| \to |X|/G$ *is the quotient map.*

Proof. We leave it as an exercise to verify that the equivalence relation generated by G-operation in $|X|$ is a cellular equivalence relation and that the conditions of II.5.9 are satisfied. Thus $|X|/G$ is a CW complex. Since each $g \in G$ is an ss isomorphism, if x is nondegenerate then $g(x)$ is also for each g in G. Thus, if a point q of $|X|/G$ has a pre-image in $|X|$ with irreducible representation (x_n,t) (t an interior point of Δ^n), then the other points of its pre-image are the points with irreducible representation $(g(x_n),t) = |g|(x_n,t)$, which are precisely the points that $|p|$ identifies with (x,t). Thus the maps $h = |p|(p')^{-1}$ and $p'(|p|)^{-1}$ are well defined and are inverses. Finally, if we restrict h or h^{-1} to a single simplex of $|X|/G$ or $|X/G|$, we see that it is a homeomorphism onto the corresponding simplex of the other space. Thus h (and h^{-1}) are regular cellular homeomorphisms. ∎

(3) **Adjunction spaces.** Let $f:A \to Y$ be an ss map where A is a subcomplex of X. The adjunction complex $Y \cup_f X$ is the quotient of the (ss) disjoint union $Y \cup X$ by the equivalence relation that identifies simplexes of A with their images under f, and whose ss operations are the quotient operations. Let $q:X \cup Y \to Y \cup_f X$ be the quotient map; it is an ss map. Note that the realization of the disjoint union $|X \cup Y|$ is the disjoint union of the realizations $|X| \cup |Y|$.

PROPOSITION **5.7.** *The map* $|q| : |X \cup Y| \to |X \cup_f Y|$ *is a quotient map and induces a homeomorphism* β *with* $|X| \cup_{|f|} |Y|$ *such that* $\beta |q| = \bar{q}$, *where* $\bar{q}$ *is the corresponding quotient map* $|X| \cup |Y| \to |X| \cup_{|f|} |Y|$.

Proof. As in the last proposition, q being surjective implies that $|q|$ is a quotient map and one must only prove that $|q|$ identifies two points if and only if $\bar{q}$ does. An argument similar to the previous one accomplishes this. ∎

COROLLARY **5.8.** *The adjunction space of a simplicial map is an ssc.* ∎

6. SIMPLICIAL SUBDIVISION OF SEMISIMPLICIAL COMPLEXES

Let X be a cell complex containing a simplicial subcomplex A. A simplicial subdivision of X rel A is a subdivision of X to a simplicial complex in which the simplexes of A are not subdivided. The object of the next few sections is to prove theorem 6.1 below, which roughly states that if X is an ssc containing the (ordered) simplicial complex A, then there is a simplicial subdivision of $|X|$ rel $|A|$. The homeomorphism between the realization $|X|$ and the simplicial complex which defines this triangulation is not the realization of an ss map, but we will show that it is homotopic to the realization of an ss map in a strong way. After stating the theorem, we will draw some conclusions and give the rather lengthy proof in sections 7, 8, and 9. In the proof of corollary 6.2 and theorem 6.7, we will use the simplicial approximation theorem for simplicial complexes, a well-known and documented result about maps of simplicial complexes, a proof of which can be found in [10].

Let $\mathcal{C}'$ be the category consisting of pairs (X,A), where X is an ssc, and A is an (ordered) simplicial subcomplex, and in which the maps are ss maps $f:(X,A)\to(Y,B)$ of pairs. Let $\mathcal{C}''$ be the category of ordered simplicial complexes and simplicial (order preserving) maps. The following is essentially due to Barratt [2].

THEOREM **6.1.** *Let (X,A) be in $\mathcal{C}'$, and let $|X|$, $|A|$ be the corresponding realizations. Then there is an ordered simplicial complex D_AX, a homeomorphism $t:|D_AX|\to|X|$, and an ss map $\lambda:D_AX\to X$ such that:*

(i) *there is a subcomplex A' of D_AX mapped isomorphically onto A by λ;*

(ii) *t defines a subdivision of the CW complex $|X|$ and on $|A'|$ reduces to $|\lambda|$;*

(iii) *$|\lambda|$ is homotopic to t by a homotopy F such that on $|A'|$, F is the constant homotopy, and for each cell $|e|$ outside of $|A'|$, the map F takes $|e|\times I$ into the smallest cell $|x|$ which contains $t(|e|)$.*

Moreover, D defines a functor from $\mathcal{C}'$ to the subcategory $\mathcal{C}''$, and λ defines a natural transformation from D to the identity.

COROLLARY **6.2.** *Let (X,A) and (Y,B) be pairs in $\mathcal{C}'$, and let $f:(|X|,|A|)\to(|Y|,|B|)$ be a map whose restriction to $|A|$ is* *a simplicial map* $|g|$. *Then* f *is homotopic to the realization of an ss map* f' *of subdivisions of* $|X|$ rel $|A|$ *and* $|Y|$ rel $|B|$, *which reduces to* $|g|$ *on* $|A|$.

If $|X|$ *and* $|Y|$ *are finite* CW *complexes (and therefore metrizable), then, for any prescribed* $\epsilon>0$, *the homotopy between* f *and its ss approximation may be chosen so that on* $|A|$ *it is the identity and does not displace any point outside of an* ϵ*-disc.*

Proof. Apply the standard simplicial approximation theorem to the map $f''\colon |D_AX| \to |D_BY|$, where $f''=t_Y^{-1}ft_X$. ∎

Our first application is to show that certain topological operations on simplicial complexes, those which can be produced by semisimplicial constructions followed by realization, lead to triangulable spaces.

Corollary **6.3.** *Let* A *be a subcomplex of the (geometric) simplicial complex* X. *Then* X/A *has a triangulation which contains the subcomplex of all simplexes of* X *that do not meet* A.

Proof. After choosing any vertex order, X and A become realizations of an ssc and (ordered) simplicial subcomplex, and X/A becomes the realization of the quotient of the two by proposition 5.4. Hence the result follows from theorem 6.1. ∎

Corollary **6.4.** *Let* X *and* Y *be simplicial complexes and* $f\colon A\to Y$ *a map of a subcomplex into* Y. *Then the adjunction space* $Y\cup_f X$ *has a simplicial subdivision containing* Y *as a subcomplex, as well as the subcomplex of* X *consisting of all simplexes that do not meet* A.

Proof. After choosing some suitable vertex orders, f becomes the realization of an ordered simplicial map of a subcomplex of one ordered simplicial complex into another. With this ss structure, the result follows from proposition 5.7. ∎

Corollary **6.5.** *If* $|X|$ *is an ordered simplicial complex and* G

is a group of ss homeomorphisms (the realizations of ss isomorphisms), then the quotient $|X|/G$ *has a simplicial subdivision.*

Proof. By proposition 5.6, $|X|/G$ is homeomorphic to the ssc $|X/G|$ which has a simplicial subdivision. ∎

COROLLARY **6.6.** *Let* K *be a simplicial complex and* G *a group of simplicial homeomorphisms. Then* K/G *has a simplicial subdivision.*

Proof. Let K' be the barycentric subdivision of K. It is a partially ordered simplicial complex in which the vertices of each simplex are linearly ordered by the dimensions of the cells of which they are barycenters, and is thus an ssc. Since each $g \in G$ is a simplicial homeomorphism, it maps barycenters into barycenters, and cells of K' into cells of K' preserving this partial order. Thus each g determines an ss map of the ssc K' to itself. The quotients K'/G and K/G are evidently homeomorphic and the former is a subdivision of the latter. By the previous corollary, K'/G has a simplicial subdivision. ∎

Our second application is the following.

THEOREM **6.7.** *The map* $j_X: |S(X)| \to X$ *induces isomorphisms of homotopy groups in all dimensions.*

Proof. Let $[f] \in \pi_n(X,*)$ be the class of the map $f:(S^n,*) \to (X,*)$. We assume that S^n is the realization of an ordered simplicial complex K with vertex $*$. Then there is a unique ss map $f': K \to S(X)$ such that $f = j_X|f'|$. Let the base point be chosen in $|S(X)|$ at a vertex lying in the pre-image $j_X^{-1}(*)$. Then f must map the vertex $*$ of $|K| = S^n$ onto $*$, since $j_X|f'|(*)$ is the vertex $*$ of X. Hence $j_{X\#}: \pi_n(|S(X)|,*) \to \pi_n(X,*)$ maps the class of $|f'|$ onto $[f]$, and therefore $j_{X\#}$ is surjective.

Suppose that $g:(S^n,*) \to (|S(X)|,*)$ composes with j_X to give a null-homotopic map. Again we assume S^n is an ordered simplicial complex K with vertex at $*$. By theorem 6.1, the map g is homotopic to an ss map $|g'|: |K| \to |S(X)|$, and thus $j_X|g'|$ is null-homotopic. There is thus a map $H: |K| \times I \to X$ which reduces to $j_X|g'|$ on $|K| \times \{0\}$, is constant at $*$ on $|K| \times \{1\}$, and sends $\{*\} \times I$ into $*$. Also $|K| \times I$ has a triangulation containing $|K|$ as the subset $|K| \times \{0\}$ and $\{*\} \times I$ as a 1-simplex. Hence there is a factorization $H = j_X|H'|$, where H' maps the ssc $K \times I$ into $S(X)$. By uniqueness of the factorization $|g'| = j_X|g'^*|$, the map $|H'|$ must reduce to $|g'|$ on $|K| \times \{0\}$. It is thus a null-homotopy of $|g'|$, and g must be null-homotopic. We see that $j_{X\#}$ is injective and is therefore an isomorphism. ▮

7. *BARYCENTRIC SUBDIVISION OF SEMISIMPLICIAL COMPLEXES*

We first introduce barycentric subdivision of ssc's. This leads to ssc's whose realizations are more simple CW complexes than is generally the case.

Definition **7.1.** Let Sd $\sum(n)$ be the ordered simplicial complex defined as follows. Partially order the faces of $\sum(n)$ by defining $\sigma < \tau$ if τ is a face of σ. Now let Sd $\sum(n)$ have vertices the faces of $\sum(n)$ and simplexes the nondecreasing linearly ordered sequences of faces of $\sum(n)$. If $\varphi: \sum(n) \to \sum(m)$ is an ordered simplicial map (monotone on vertices), then Sd φ maps $\langle \sigma_0, \cdots, \sigma_n \rangle$ into $\langle \varphi\sigma_0, \cdots, \varphi\sigma_n \rangle$. In particular, this defines

$$\text{Sd}\, d_i^*: \text{Sd} \textstyle\sum(n) \to \text{Sd} \sum(n+1),$$

and

$$\text{Sd}\, s_i^*: \text{Sd} \textstyle\sum(n+1) \to \text{Sd} \sum(n).$$

If X is an ssc, let $M'(X)$ be the union $\bigcup$ Sd $\sum(n) \times x_n$ of one copy of Sd $\sum(n)$ for each n-simplex x_n of X, given the obvious semisimplicial structure. We define an equivalence relation $\mathfrak{R}$ on $M'(X)$ by setting $\sigma \times d_i x_n \sim \text{Sd}\, d_i^* \sigma \times x_n$ for each simplex $\sigma \in \sum(n-1)$, and $\sigma \times s_i x_n \sim \text{Sd}\, s_i^* \sigma \times x_n$ for each $\sigma \in \sum(n+1)$.

Definition **7.2.** Let Sd X be the ssc $M'(X)/\mathcal{R}$. The ss operations on Sd X are thus given by: $d_i(\sigma,x_n) = (d_i\sigma,x_n)$ and $s_j(\sigma,x_n) = (s_j\sigma,x_n)$, where (σ,x_n) denotes the equivalence class of the element $\sigma \times x_n$ of Sd $\sum(n) \times x_n$.

If $f:X \to Y$ is an ss map, then $f(\sigma,x_n) = (\sigma,fx_n)$.

It is easy to verify that these definitions define a functor from the category of ssc's to itself.

To describe further the structure of Sd X, we want to describe a unique representative from each equivalence class of the quotient set Sd X. We note that any simplex σ of Sd $\sum(n)$ has a unique representation Sd $F^*\tau$, where $F^*: \sum(p) \to \sum(n)$ is a face operator embedding $\sum(p)$ as the F face of $\sum(n)$ and τ is an *interior* simplex of Sd $\sum(p)$, that is, has for 0th vertex the vertex corresponding to the simplex $\sum(p)$ itself. (Upon realization, such a simplex of Sd Δ^p would contain interior points of Δ^p.) This representation is determined by finding the unique smallest face $F^*\sum(p)$ "containing σ," the one whose vertices are the distinct vertices from among the sequence of faces that give σ in Sd $\sum(n)$. The face operator F^* is the one that embeds this face in $\sum(n)$. The following lemma is proved exactly as was lemma 4.1.

LEMMA **7.3.** *In the equivalence class of every element $\sigma \times x_n$ of $M'(X)$ there is a unique irreducible representative $\tau \times y$, where τ is interior to* Sd $\sum(p)$ *and y_p is nondegenerate. We define the selection function θ that picks this class out of the given class as follows: represent $\sigma \times x_n$ uniquely as* Sd $F^*\tau_1 \times x_n$, *where F^* is a face operator and τ_1 interior; let $Fx_n = Dy_p$, where D is a degeneracy operator and y_p is nondegenerate, and let $\tau =$* Sd $D^*\tau_1$. *Then $\theta(\sigma \times x_n) = \tau \times y_p$.* ∎

Thus, as in 4.5, we have the following result.

LEMMA **7.4.** *The characteristic ss map φ of the simplex (σ,x_n) of* Sd X *is the composition of the inclusion map of σ into* Sd $\sum(n)$ *with* Sd φ_x, *where φ_x is the characteristic map $\varphi_x: \sum(n) \to X$ of x_n. Also,* Sd φ_x *is bijective on the interior simplexes of* Sd $\sum(n)$. ∎

PROPOSITION **7.5.** *For any ssc X, there is a homeomorphism $t\colon |\operatorname{Sd} X| \to |X|$ identifying $|\operatorname{Sd} X|$ with a subdivision of the CW complex $|X|$.*

Proof. We will subdivide $|X|$ by a modified "star-subdivision" process. In each face of each simplex of $M(X)$ we will choose an interior point. We proceed inductively as follows. If $\sigma \times x_n$ is a vertex, we select it itself as the interior point. Suppose that we have selected one interior point of each simplex of dimension less than n in $M(X)$ so that, whenever the face $\sigma \times x_n$ of $\Delta^n \times x_n$ is identified with the face $\tau \times x_m'$, the points chosen are identified with each other. Let $\sigma \times x_m$ be an n-dimensional face of $\Delta^m \times x_m$. If $\sigma = F^* \tau^n$ and Fx_m is a nondegenerate n-simplex of X, choose the barycenter of this face. If not, say $Fx_m = Dx_p'$, where D is a degeneracy of the nondegenerate p-simplex x_p' for $p < n$; then choose any interior point of σ that is identified with the point chosen inductively in $\Delta^p \times x_p'$. It is easy to see that this extends the selection of interior points over dimension n, and thus we can select such "pseudo-barycenters" in each simplex of $M(X)$.

We now subdivide each cell $|x_n|$ of $|X|$ by starring $\Delta^n \times x_n$ for each simplex x_n of X and joining the already defined subdivision on the boundary of each face to the chosen interior point by straight line segments, portions of planes, etc. Because the maps d^* and s^* which identify the simplexes of $M(X)$ to give $|X|$ are linear, they will define a consistent subdivision of $|X|$. Clearly this subdivision is the homeomorphic image of $|\operatorname{Sd} X|$. We define such a homeomorphism as follows. Take any irreducible representative $\sigma \times x_n$ of a simplex and map it into $\Delta^n \times x$ in $M(X)$ by mapping its vertices onto the "pseudo-barycenters" to which they correspond in $\Delta^n \times x_n$, the points already chosen, and extending linearly. It is easy to check that this is indeed a homeomorphism. ∎

PROPOSITION **7.6.** *There is a natural ss map $\lambda_X\colon \operatorname{Sd} X \to X$ and a homotopy $F\colon |\operatorname{Sd} X| \times I \to |X|$ between $|\lambda_X|$ and t for any map t as in the preceding proposition, with the further property that if $|x_n|$ is an n-cell of $|\operatorname{Sd} X|$, then F maps $|x_n| \times I$ into the cell whose interior contains $t(|x_n|)$.*

Proof. The map λ_X is the one that sends the simplex $\sigma \times x_n$ into $\varphi_x(\tau)$, where τ is the simplex of $\sum(n)$ whose vertices are the last vertices of σ. It is easy to check that this is well defined, and it maps a simplex $\sigma \times x_n$ of Sd X into a face of x_n. The homotopy F is defined in the simplexes of $M(X) = \bigcup \Delta^n \times X_n$ and then projected onto $|X|$; it is the linear homotopy which moves each "pseudo-barycenter" toward the last vertex of the simplex to which it corresponds. ∎

If A is an ordered simplicial subcomplex of X, we modify these constructions to get a subdivision in which only the cells outside $|A|$ are divided.

Let A be a subcomplex of $\sum(n)$. Then $\mathrm{Sd}_A \sum(n)$ is the ordered simplicial complex whose vertices are those of A and all simplexes of $\sum(n)$ lying outside A. A simplex of this complex is given by $\langle v_0, \cdots, v_k, v_{k+1}, \cdots, v_n \rangle$, where $v_0, \cdots, v_k$ are simplexes of X not in A, $v_{k+1}, \cdots, v_n$ are vertices of A with $v_{k+1} \leq \cdots \leq v_n$ in the order on A, and all are contained in v_k. If $f: (\sum(n),A) \to (\sum(m),B)$, where A and B are subcomplexes of $\sum(n)$ and $\sum(m)$, respectively, then $\mathrm{Sd}' f$: $\mathrm{Sd}_A \sum(n) \to \mathrm{Sd}_B \sum(m)$ sends $\langle v_0, \cdots, v_n \rangle$ into $\langle fv_0, \cdots, fv_n \rangle$.

We modify the construction of $M'(X)$ as follows. Let $M_A'(X) = \bigcup \mathrm{Sd}_{[A:x]} \sum(n) \times x_n$, where for each n-simplex x_n with characteristic map φ_x, we set $[A:x] = \varphi_x^{-1}(A) \subset \sum(n)$. We note that if $x_n = d_i y_{n+1}$, then d_i^* induces a map $\mathrm{Sd}' d_i^*$: $\mathrm{Sd}_{[A:x]} \sum(n) \to \mathrm{Sd}_{[A:y]} \sum(n+1)$, and, similarly, if $x_n = s_j y_{n-1}$, then s_j^* induces a map $\mathrm{Sd}' s_j^*$: $\mathrm{Sd}_{[A:x]} \sum(n) \to \mathrm{Sd}_{[A:y]} \sum(n-1)$. We define the relation $\mathcal{R}'$ on $M_A'(X)$ using these maps instead of Sd d_i^* and Sd s_j^*.

Definition **7.7.** Let $\mathrm{Sd}_A X$ be the ssc $M_A'(X)/\mathcal{R}'$. The ss operations are given by $d_i(\sigma,x_n) = (d_i\sigma,x_n)$ and $s_j(\sigma,x_n) = (s_j\sigma,x_n)$, where (σ,x_n) represents the equivalence class of the element $\sigma \times x_n$ of $\mathrm{Sd}_{[A:x]} \sum(n) \times x_n$.

If $f: (X,A) \to (Y,B)$ is an ss map, then $f(\sigma,x) = (\sigma, f(x))$.

It is easy to see that the analogues of 7.3 and 7.4 are valid and that we can modify the construction of t, in propositions 7.5 and

7.6, so that it will give a subdivision in which only the cells of $|X|$ outside $|A|$ are divided.

In this case, we define λ: $\mathrm{Sd}_A X \to X$ as follows. If $\sigma \in \mathrm{Sd}_{[A:x]} \sum(n)$, $\lambda(\sigma)$ is the simplex of $\sum(n)$ that we obtain by assigning the simplex that has the vertices of A of σ and whose other vertices, corresponding to faces of $\sum(n)$ outside A, are the last vertices of these faces of $\sum(n)$. It is easy to check that this defines a mapping $\mathrm{Sd}_A X \to X$ by setting $\lambda(\sigma,x) = \varphi_x(\lambda(\sigma))$, for each point (σ,x). The set of classes of the form (σ,x) for $x \in A$ is an ss subcomplex A' of $\mathrm{Sd}_A X$ isomorphic to A. Note that $\mathrm{Sd}_{[A:x]} \sum(n) = \sum(n)$ if $x \in A$, and λ is an isomorphism of A' onto A.

We can summarize our results about this relative barycentric subdivision as follows, in a form parallel to that of theorem 6.1.

THEOREM **7.8.** *Relative subdivision* Sd′ *is a functor from the category of ss pairs and maps into itself. There is a transformation of functors* λ' *from* Sd′ *to the identity on this category, and for each* (X,A) *a homeomorphism* t: $|\mathrm{Sd}_A X| \to |X|$ *such that:*

(i) λ' *maps a subcomplex* A' *of* $\mathrm{Sd}_A X$ *isomorphically onto* A;

(ii) t *defines a subdivision of the CW complex* $|X|$ *and reduces to* $|\lambda'|$ *on* $|A'|$;

(iii) t *is homotopic to* $|\lambda'|$ *via a homotopy constant on* $|A'|$ *which for each cell* $|e|$ *of* $|\mathrm{Sd}_A X|$, *maps* $|e| \times I$ *into the smallest cell which contains* $t(|e|)$. ∎

8. REGULATED SEMISIMPLICIAL COMPLEXES

Let x be a simplex of the ssc X and let φ: $\sum(n) \to X$ be its characteristic map. We say that x is *regulated* if, for each pair of faces $y \neq y'$ where y contains the 0th vertex, $\varphi(y) \neq \varphi(y')$. If each nondegenerate simplex of X is regulated, we say that X is regulated.

PROPOSITION **8.1.** *For any ssc* X, *the ssc* $\mathrm{Sd}\, X$ *is regulated.*

Proof. It follows from lemma 7.4 that for any nondegenerate x in X, with characteristic map φ, the map Sd φ: Sd $\sum(n) \to$ Sd X is bijective on the "interior simplexes," those simplexes of Sd $\sum(n)$ which have for the 0th vertex the vertex corresponding to Sd $\sum(n)$. Moreover, no such interior simplex is identified with one on the boundary. ▮

The property of being regulated is the only property of Sd X that we will need.

Let x be a regulated n-simplex of X, let φ_x: $\sum(n) \to X$ be its characteristic map, and $|\varphi_x|$ the corresponding realization, a characteristic map for the n-cell $|x|$ of $|X|$. We will first analyze the identifications of points of Δ^n under the map $|\varphi_x|$.

Lemma 8.2. *There is an integer p and a face operator F such that $|\varphi|$ is bijective on all open cells outside of the p-dimensional face $F^*\Delta^p$ of Δ^n. On this face, there is a face $F'^*\Delta^q$ such that the restriction of $|\varphi_x|$ to $F^*\Delta^p$ is $|\varphi_{F'x}|D^*$, where D is a suitable degeneracy map and $F'x$ a nondegenerate face of x.*

Proof. We note that if $|\varphi_x|$ identifies interior points of two open cells, then it identifies all interior points of one with the other. Thus no interior point of the n-dimensional simplex Δ^n is identified with any other point. Let p be the highest dimension in which an interior point of a p-dimensional cell, say $F^*\Delta^p$, is identified with an interior point of another cell, say $F''^*\Delta^r$ with $p \geq r$. Then clearly φ_x identifies $F^*\sum(p)$ with a degeneracy of $F''^*\sum(r)$. Let $F'''^*\sum(s)$ be the smallest face of $\sum(n)$ containing both $F^*(\sum(p))$ and $F''^*\sum(r)$, the face whose set of vertices is the union of the sets of vertices of the two faces. If we assume that $F''^*\sum(r)$ is not a face of $F^*\sum(p)$, then the set of vertices is strictly bigger than that of $F^*\sum(p)$; hence $F'''^*\sum(s)$ is a strictly larger face of $F^*\sum(p)$. This means in particular that $F'''x = \varphi_x(F'''^*\sum(s))$ is nondegenerate since its dimension is larger than p. Because we are assuming x is a regulated simplex and because $F^*\sum(p)$ and $F''^*\sum(r)$ are simplexes that are identified (or whose degeneracies are identified) by φ_x, it follows that neither of them can have the 0th vertex of $F'''^* \sum(s)$ for a vertex. This contradicts the hypothesis that $F'''^* \sum(s)$ is the smallest simplex whose vertices contain those of the two given faces. Hence if x is regulated, it is not possible for $F''^* \Delta^r$ to lie outside of $F^* \Delta^p$.

Suppose that t is an interior point of Δ^p and that (F^*t,x) is the corresponding point of the simplex $\Delta^n \times x$ in the union $M(X)$. Suppose also that $|\varphi_x|$ identifies this point with (F'^*u,x), where $F'x$ is a nondegenerate q-face of x and u an interior point of Δ^q. That is, $(F^*t,x) \sim (u,F'x)$, and the latter is an irreducible point of $M(X)$. Then the selection function θ sends (F^*t,x) into $(u,F'x)$, so that $Fx = DF'x$ and $D^*t = u$, where D is a suitable degeneracy operator, by definition of θ. It thus follows that for any other point of $F^*\Delta^p$, we have $|\varphi_x| (F^*t',x) = |\varphi_{F'x}| (D^*t',Fx)$, and therefore the restriction of $|\varphi_x|$ to the face $F^*\Delta^p$ is the composition $|\varphi_{F'x}| D^*$. ∎

To find the remaining identifications produced by $|\varphi_x|$, repeat this analysis on (Δ^q,F'^*x). Since $F'x$ is a nondegenerate face of a regulated simplex, it is regulated. Thus there is a p_2-face of $\Delta^q = (\Delta^q,F'^*x)$ such that, on each open cell of dimension higher than p_2 and on each open cell of dimension p_2 outside this face, the map $|\varphi_{F'x}| = |\varphi_x| \cdot F'^*$ is bijective, etc. Iteration of the preceding lemma gives the following corollary.

Corollary 8.3. *If x is a regulated n-simplex of X, then $|\varphi_x| : \Delta^n \to |X|$ makes the following identifications (and no others). There is a descending sequence of faces of Δ^n:*

$$\Delta^n = \tau_0 \supset \sigma_1 \supset \tau_1 \supset \sigma_2 \supset \tau_2 \supset \cdots \supset \sigma_r \supset \tau_r$$

of dimensions: $\dim \sigma_i = p_i$, $\dim \tau_i = q_i$, *and degeneracy operators D_i such that:*

(i) $|\varphi_x| \,|\, \tau_i$ *is bijective on all open cells outside of* σ_{i+1};

(ii) $|\varphi_x| \,|\, \sigma_{i+1} = (|\varphi_x| \,|\, \tau_{i+1}) D^*_{i+1}$;

(iii) $|\varphi_x|$ *is bijective on the interior of* τ_{i+1}. ∎

We will now prove that a regulated n-simplex realizes to a regular n-cell in $|X|$.

LEMMA 8.4. *Let $\Delta^n \supset \sigma \supset \tau$ be proper faces, let $D^*: \sigma \rightarrow \tau$ be a degeneration map, let L be the quotient of Δ^n by the identifications of D^*, and let $\varphi: \Delta^n \rightarrow L$ be the quotient map. Then there is a homeomorphism $h: \Delta^n \rightarrow L$ such that if $r \leq q = \dim \tau$, we have $hi = \varphi \mid \xi$ for any r-dimensional face ξ with inclusion map $i: \xi \rightarrow \tau$.*

Proof. Let σ' be the face of Δ^n opposite σ. Each point P of Δ^n has a unique representation $P = (1-t)Q + tQ'$, where $Q \in \sigma$, $Q' \in \sigma'$. Define $\rho(P)$ to be P if $1/2 \leq t \leq 1$; and $t(Q+Q') + (1-2t)D^*(Q)$ if $0 \leq t \leq 1/2$. Then:

(i) $\rho(P) = \rho(P')$ if and only if $\varphi(P) = \varphi(P')$;
(ii) Im ρ is compact a convex subset of Δ^n; and
(iii) Im ρ contains the simplex ω of Δ^n spanned by σ' and τ, and the subset of Δ^n consisting of points with $t \geq 1/2$.

We leave verification of these facts to the reader. From (i) and the compactness of Δ^n, it follows that Im ρ is homeomorphic to L. From (ii) and (iii), Im ρ contains a small copy of Δ^n, the convex set with vertices $\{v_i'\}$, corresponding to the vertices of σ', and $\{(v_i + b')/2\}$, where b' is the barycenter of σ' and v_i is a vertex of σ'. By radial projection from the barycenter of the face of this simplex determined by the simplexes corresponding to σ and τ, we get a homeomorphism $h: \Delta^n \rightarrow$ Im ρ which restricts to a linear homeomorphism of the simplex corresponding to ω onto ω. ∎

PROPOSITION 8.5. *Let x be a regulated n-simplex of the ssc X. Then $|x|$ is a regular n-cell of the CW complex $|X|$. Hence if X is a regulated ssc, then $|X|$ is a regular CW complex.*

Proof. Suppose x is the simplex for which φ_x has the form described in corollary 8.3. Let L_i be the quotient of Δ^n by the identifications of $D_i^*, \cdots, D_i^*$, and let $\varphi_i: \Delta^n \rightarrow L_i$ be the quotient map for each i. Suppose for some k we have a homeomorphism $h_k: \Delta^n \rightarrow L_k$ such that $\varphi_k \mid \tau_{k+1} = h_k \mid \tau_{k+1}$. This is true when $k=1$ by 8.3. Then the map induced by D^*_{k+1} on $h_k(\tau_{k+1}) \subset L_k$ corresponds to D^*_{k+1} on $\tau_{k+1} \subset \Delta^n$. By 8.3, we have a homeomorphism $\bar{h}: \Delta^n \to L_{k+1}$ such that if $\bar{\varphi}: L_k \to L_{k+1}$ is the quotient map, then $\bar{h} \mid \tau_{k+2} = \bar{\varphi} \mid \tau_{k+2}$. Set $h_{k+1} = \bar{h} \bar{h}_k$ and $\varphi_{k+1} = \bar{\varphi}\varphi_k$. After r steps, we arrive at a homeomorphism $h_r: \Delta^n \to L_r = \mid x \mid$. ∎

9. *THE FUNCTOR* $*$

Let X be an ssc. We define an ordered simplicial complex $*X$ as follows. The *vertices* of $*X$ are the nondegenerate simplexes of X ordered by: $x < y$ if y is a face of x. A sequence of vertices $\langle x^0, \cdots, x^n \rangle$ for which $x^0 \leq \cdots \leq x^n$ is a simplex of $*X$. If $f: X \to Y$ is a ss map, we define an order-preserving simplicial map $*f: *X \to *Y$ by letting $*f(\langle x \rangle)$ be the nondegenerate simplex of Y, of which $f(x)$ is a degeneracy, and extending to a simplicial map.

This defines $*$ as a functor from the category of ssc's to the subcategory of ordered simplicial complexes. On the simplicial subcategory, this is just barycentric subdivision.

Looking back at the description of star subdivision of regular CW complexes, we see that if $\mid X \mid$ is a regular CW complex, this is the starring simplicial subdivision. In this case, we note that the simplex $\langle x^0, \cdots, x^n \rangle$ corresponds to a simplex lying in the cell $\mid x^0 \mid$ of $\mid X \mid$; and if $n = \dim x^0$, this will correspond to a simplex whose interior lies in the interior of $\mid x^0 \mid$.

Let $t_X: \mid *X \mid \to \mid X \mid$ be a triangulation of the regular CW complex $\mid X \mid$ by star subdivision.

PROPOSITION 9.1. *Let* $\mid X \mid$ *be a regular CW complex, and let* $\varphi: *X \to X$ *be an ss map such that* $\varphi(\langle x^0, \cdots, x^n \rangle)$ *is a face of* x^0 *for each simplex nondegenerate in* $*X$. *Then* $\mid \varphi \mid$ *is homotopic to* t_X *by a homotopy which maps* $\mid \langle x^0, \cdots, x^n \rangle \mid \times I$ *into the cell whose interior contains the interior of* $t_X(\langle x^0, \cdots, x^n \rangle)$.

Proof. This is obvious, for the images of each simplex of $\mid *X \mid$ under $\mid \varphi \mid$ and t_X are contained in the same contractible subset of $\mid X \mid$, the cell whose interior contains the interior of the t_X

image. Thus the assignment of this cell is an aspherical carrier carrying both maps t_X and $|\varphi|$. ▮

PROPOSITION 9.2. *Let X be a regulated ssc. Then there is an ss map $*\lambda: *X \to X$ natural with respect to maps of regulated ssc's such that $*\lambda(\langle x^0, \cdots, x^n \rangle)$ is a face of x^0 for each nondegenerate simplex of $*X$.*

Proof. Let $\tau = \langle x^0, \cdots, x^n \rangle$ be an n-simplex of $*X$. Let $\varphi: \sum(m) \to X$ be the characteristic map of a nondegenerate simplex $x^0 \supset x$. Let σ^m be the fundamental simplex $\langle 0, \cdots, m \rangle$ of $\sum(m)$, and let $\{F_i\sigma\}$ be a descending sequence of faces of σ^m for which $\varphi(F_i\sigma^m) = F_ix = x^i$. Let $L(i)$ be the last vertex of $F_i\sigma$, and let $L = \langle L(n), \cdots, L(0) \rangle$ be an ordered simplex of $\sum(m)$ and the face $F\sigma$ of σ. Define $*\lambda(\tau) = \varphi(L) = Fx$. If we can show that $*\lambda(\tau)$ is well defined, then clearly $*\lambda(\langle x^0, \cdots, x^n \rangle)$ is a face of x^0 and $*\lambda$ is a natural ss map.

Suppose that $F_i'\sigma$ is a second descending sequence of faces of $\sum(m)$ such that $\varphi(F_i\sigma) = x^i$. Let $L' = \langle L'(n), \cdots, L'(0) \rangle$ be the corresponding simplex of last vertices. We must prove that $\varphi(L) = \varphi(L')$.

The proof will be by induction on n. If $n = 0$, there is clearly nothing to prove. We consider separately the case $n = 1$. Here, let $x^0 = \varphi\langle v_0, \cdots, v_m \rangle = \varphi\langle w_0, \cdots, w_m \rangle$ and $x^1 = \varphi\langle v_{j_0}, \cdots, v_{j_k} \rangle = \varphi\langle w_{i_0}, \cdots, w_{i_k} \rangle$. Let $z = \varphi\langle v_{j_k}, v_m \rangle$ and $z' = \varphi\langle w_{i_k}, w_m \rangle$. We will prove $\varphi\langle v_{j_k}, v_m \rangle = \varphi\langle w_{i_k}, w_m \rangle$. Since $\varphi\langle v_0, \cdots, v_m \rangle = \varphi\langle w_0, \cdots, w_m \rangle$, the map $h: \langle v_0, \cdots, v_m \rangle \to \langle w_0, \cdots, w_m \rangle$ mapping v_i into w_i is an ss isomorphism of the simplex of $\langle v_0, \cdots, v_m \rangle$ with that of $\langle w_0, \cdots, w_m \rangle$, and $\varphi \mid \langle v_0, \cdots, v_m \rangle = (\varphi \mid \langle w_0, \cdots, w_m \rangle) h$. Thus $\varphi\langle w_{i_k}, w_m \rangle = \varphi\langle v_{i_k}, v_m \rangle$ and it suffices to prove $\varphi\langle w_{j_k}, w_m \rangle = \varphi\langle w_{i_k}, w_m \rangle$, where we have the following information: $\varphi(w_{i_k}) = \varphi(w_{j_k})$, and w_{i_k}, w_{j_k}, w_m are mapped onto vertices of a regulated simplex. We may assume $i_k \leq j_k$. Consider $\varphi\langle w_{i_k}, w_{j_k}, w_m \rangle = z''$. Then since its zero-th and first vertices are identified, and z'' is a face of a regulated simplex, it is degenerate. If $z'' = s_0 z^*$, then $d_0 z'' = d_1 z''$ and $\varphi\langle w_{j_k}, w_m \rangle = \varphi\langle w_{i_k}, w_m \rangle$. If $z'' = s_1 z^*$, then $d_1 z'' = d_2 z''$, and $\varphi\langle w_{i_k}, w_m \rangle = \varphi\langle w_{i_k}, w_{j_k} \rangle$ is again degenerate. Thus $d_0\varphi\langle w_{i_k}, w_m \rangle = d_1\varphi\langle w_{i_k}, w_m \rangle$, so that $\varphi(w_{i_k}) = \varphi(w_{j_k}) = \varphi(w_m)$. Then $\varphi\langle w_{j_k}, w_m \rangle$ is also a degeneracy of $\varphi\langle w_m \rangle$, and again $z = z'$.

Suppose inductively that $*\lambda$ is well defined on all sequences of faces with fewer than $n+1$ elements and let τ be one with $n+1$ vertices.

Let $z=\varphi(L)$, $z'=\varphi(L')$. Note that for any proper face operator F we have $Fz=Fz'$. Regarding F to be a face operator on the ordered simplicial complex $\sum(n)$, and thus a deletion operator, when we delete the indicated vertices from L and L' we get the simplexes whose vertices are the last vertices for the sequence of faces obtained from the given $\langle x^0, \cdots, x^n\rangle$ by performing the indicated deletion operation. Hence we can apply the induction hypothesis and conclude that, for any proper face operator F,

$$Fz=\varphi(FL)=\varphi(FL')=Fz'.$$

Suppose first that z and z' are degenerate, say $z=Dz_1$, $z'=D'z_1'$, where D and D' are degeneracy operators and z_1, z_1' nondegenerate. Let F and F' be face operators such that $FD=F'D'=1$. Then $z_1=FD'z_1'=FD'F'Dz_1$, so that $FD'F'D=1$. Correspondingly, $F'DFD'=1$, and thus $FD'=F'D=1$ and $z_1=z_1'$. It is easy to see that $z=z'$.

Next suppose that z is degenerate, say $z=s_iz_i$. Then $d_iz=d_{i+1}z$. Hence $d_iz'=d_iz=d_{i+1}z=d_{i+1}z'$ by the induction hypothesis. Thus z' is degenerate because x is regulated, and we are back in the case just above.

Finally, suppose that both z and z' are nondegenerate. Our induction hypothesis implies that all corresponding faces of z and z' are equal. The result then follows from the next lemma. ∎

LEMMA **9.3.** *Let X be a regulated ssc, let x and y be nondegenerate faces of a nondegenerate simplex w of X. If $d_nx=d_ny$, then either $x=y$ or there is a nondegenerate face w_1 of dimension $n+1$ which has x and y for faces. If also $d_{n-1}x=d_{n-1}y$ and $n\geq 2$, then $x=y$.*

Proof. Let $\varphi_w\colon \sum(m)\to X$ be the characteristic map of w, and let σ be a face of $\sum(m)$ of minimal dimension such that $w_1=\varphi_w(\sigma)$ has both x and y for faces. Then w_1 is nondegenerate, for if $w_1=D\bar{w}$, where $\bar{w}$ is nondegenerate and D is a degeneracy operator, the nondegenerate faces x and y of w_1 will be faces of $\bar{w}$.

Suppose $Fw_1 = x$ and $F'w_1 = y$, where F and F' are face operators. Consider $\sigma = \langle v_{i_0}, \cdots, v_{i_r} \rangle$, a face of the fundamental simplex $\langle 0, \cdots, m \rangle$ of $\sum(m)$. Then F and F' operate as deletion operators on σ, and φ_w maps $d_n F\langle v_{i_0}, \cdots, v_{i_r} \rangle$ and $d_n F'\langle v_{i_0}, \cdots, v_{i_r} \rangle$ into the same element of X. Now either one or both of $F\sigma$ and $F'\sigma$ contain the zero-th vertex of σ, since it is a minimal simplex having $Fw_1 = x$ and $F'w_1 = y$ for faces. Thus one or both of $d_n F\sigma$ and $d_n F'\sigma$ contain v_{i_0} and have the same image under φ_w, a characteristic map of a nondegenerate and thus regulated simplex of X. By definition of a regulated ssc, $d_n F\sigma$ must be equal to $d_n F'\sigma$. Now $F\sigma$ and $F'\sigma$ simply delete all but $(n+1)$-vertices of σ; thus the vertices remaining after applying F and F' must be the same except for the last vertices in $F\sigma$, $F'\sigma$. Since σ is a minimal simplex whose image contains x and y, it must therefore be an $(n+1)$-simplex and F, F' must simply delete either the nth or the $(n+1)$st vertex. Otherwise we could delete from σ all the vertices except the common $n-1$ vertices of $F\sigma$ and $F'\sigma$ and the remaining last vertices of $F\sigma$ and $F'\sigma$.

If $n \geq 2$ and $d_{n-1}x = d_{n-1}y$, then $d_{n-1}F\sigma$ and $d_{n-1}F'\sigma$ are faces, one of which contains the 0th vertex of σ, and have the same image under φ_w. Thus they are identical because w_1 is regulated. Therefore the last vertices of $F\sigma$ and $F'\sigma$ are also equal and therefore $F\sigma = F'\sigma$. Thus $x = Fw_1 = F'w_1 = y$. ∎

From 7.6, 9.1, and 9.2, we deduce the following statement.

PROPOSITION **9.4.** *Let $\mathcal{C}'$ be the category of regulated ssc's and ss maps. Then there is a natural transformation $*\lambda$ from $*$ to 1 on $\mathcal{C}'$. Moreover, if X is any regulated ssc and $t_X\colon |*X| \to |X|$ a starring triangulation, then there is a homotopy $F\colon |*X| \times I \to |X|$ such that $F \mid (|*X| \times \{0\}) = |*\lambda_X|$, and $F \mid (|*X| \times \{1\}) = t_X$; and for each simplex σ of $*X$, $F \mid \sigma \times I$ has its image in the cell whose interior contains $t_X(\sigma)$.* ∎

If A is a simplicial subcomplex of X, we define the relative starring of X, denoted by $*_A X$, as follows. The vertices of this ordered simplicial complex are all vertices of A together with one vertex x_n corresponding to each simplex of X outside A. The ordering on these vertices is as follows: $v<v'$ if both are vertices of A and v precedes v' in the ordering on A, or, if not both are vertices of A, if v' is a face of v. A simplex is a sequence $\langle x^0, \cdots, x^n\rangle$ of vertices in which the vertices not in A satisfy $x^0 \leq \cdots \leq x^p$, and the remaining vertices of A form a simplex in A which is a face of each x^i outside A. If the pairs (X,A) and (Y,B) are regulated ssc's and simplicial subcomplexes and $f\colon (X,A) \to (Y,B)$ is an ss map, then $*f\colon *_A X \to *_B Y$ sends $\langle x^0, \cdots, x^n\rangle$ into $\langle fx^0, \cdots, fx^n\rangle$.

We define $*\lambda'\colon *_A X \to X$ as follows. If $\langle x^0, \cdots, x^n\rangle$ is a simplex of $*_A X$ with $x^0, \cdots, x^p$ the vertices corresponding to simplexes outside A, the $*\lambda'$ sends this into $\varphi_x\langle L(0), \cdots, L(n)\rangle$, where $L(i)$ is the last vertex of x^i. A modification of 9.2 shows that this is well defined, and it clearly sends a subcomplex A' (generated by the vertices of A) isomorphically onto A.

We thus arrive at the following variation of proposition 9.2.

Proposition **9.5.** *Let $\mathcal{C}''$ be the category of pairs consisting of a regulated ssc and a simplicial subcomplex. Then relative starring is a functor from $\mathcal{C}''$ to the subcategory of pairs of ordered simplicial complexes and ss maps. There is a transformation of functors $*\lambda'$ from relative starring to the identity, and for each pair in $\mathcal{C}''$ there is a homeomorphism $t'\colon |*_A X| \to |X|$ such that:*

(i) *$*\lambda'$ maps a subcomplex A' isomorphically onto A;*

(ii) *t defines a subdivision of the CW complex $|X|$ and reduces to $|*\lambda'|$ on $|A'|$;*

(iii) *t' is homotopic to $|*\lambda'|$ via a homotopy which is constant on $|A'|$ and maps $|e| \times I$ into the smallest cell of $|X|$ that contains $t(|e|)$, for each cell $|e|$ of $|*_A X|$.* ∎

This proposition together with the corresponding proposition about (relative) barycentric subdivision implies theorem 6.1. The functor D is the composition of Sd′ and relative starring; the compositions of the maps t and λ for the two cases give the t and λ called for in the theorem and, similarly, the two homotopies for the cases of Sd and $*$ give the desired homotopy between the compositions t and $|\lambda|$.

CHAPTER IV

HOMOTOPY TYPE OF CW COMPLEXES

In this chapter we will study conditions on a space that will cause it to have the homotopy type of a CW complex and prove a theorem of J. H. C. Whitehead which gives a criterion that a map between two such spaces be a homotopy equivalence. We note in passing that there is a method using elaborate algebraic techniques (the analysis of Postnikov systems) for determining, in principle, whether two such spaces have the same homotopy type, but this is beyond the scope of this book.

In a sense, the topology of a CW complex is simple because the space is put together simply, and this simplicity is often reflected in topological invariants of homotopy type that can be described algebraically. One immediately accessible example is the fact that for such a space the path components are the same as the components by II.6.8. Thus the closure of the graph of $y = \sin(1/x)$, $0 < x < 1$, is not homotopy equivalent to a CW complex. The general fact for which this is a special case can best be expressed by saying that two different cohomology theories (Alexander–Spanier and singular) must agree on a space which has the homotopy type of a CW complex.

An extension of the problem of whether or not a space is homotopy equivalent to a CW complex is whether or not an n-ad $\boldsymbol{X}$ has the homotopy type of a CW n-ad. As we shall see, the major theorems about CW homotopy type generalize simply to ones about n-ads.

Much of the material of this chapter is due to Whitehead [40, 42] and Milnor [29].

1. HOMOTOPY EQUIVALENCE AND DEFORMATION RETRACTION

We refer the reader to the introductory material (0.3) for the definitions of retraction, deformation, etc.

In the following paragraphs we relate homotopy equivalence and deformation retraction, as described subsequently in theorem 1.6. The problem of whether or not a given map is a homotopy equivalence is thus equivalent to that of whether or not a certain space is a deformation retract of another.

These remain true when spaces are replaced by n-ads, if we understand maps, homotopies, etc., of n-ads as in the introductory material (0.3), together with the following definition.

Definition **1.1.** If $\boldsymbol{f}:\boldsymbol{X}\rightarrow\boldsymbol{Y}$ is a map of n-ads, the *mapping cylinder* of $\boldsymbol{f}$ is the n-ad $\boldsymbol{M}_{\boldsymbol{f}}$ consisting of the mapping cylinder of f and the $n-1$ subspaces, the mapping cylinders of the restrictions of f.

In this section, the proofs in the case of n-ads are so similar to those in the case of a single space that we omit them.

Let $f:X\rightarrow Y$ be a map and M_f its mapping cylinder. We denote by $[x,t]$ the point of the mapping cylinder which is the image of the point $(x,t)\in X\times I$. We also identify X and Y with the corresponding subspaces of M_f. Thus $[x,1]=f(x)$ and $[x,0]=x$.

LEMMA **1.2.** *If X and Y are spaces and $f:X\rightarrow Y$ is a map, then Y is a strong deformation retract of the mapping cylinder M_f.*

Proof. Define $F:M_f\times I\rightarrow M_f$ by $F([x,t],u)=[x,u+(1-u)t]$ and $F(y,u)=y$ for $y\in Y$. One easily checks that F is a strong deformation of M_f onto Y. ∎

LEMMA **1.3.** *A subspace Y of X is a deformation retract if and only if there is a retraction r of X to Y and a deformation D of X into Y.*

Proof. If D is a deformation retraction, then it is first of all a deformation of X into Y, and $D\mid X\times\{1\}$ is a retraction of X to Y.

Conversely, let $D':X \times I \to X$ be the map such that $D'(x,t) = D(x,2t)$ for $0 \leq t \leq 1/2$ and $D'(x,t) = rD(x,2-2t)$ for $1/2 \leq t \leq 1$. Since $D(x,1) \in Y$ and r is a retraction of X to Y, the two definitions of $D'(x,1/2)$ are consistent, and therefore D' is a continuous map of $X \times I$ into X. When $t=0$, $D'(x,0) = D(x,0) = x$ since D is a deformation, and when $t=1$, $D'(x,1) = rD(x,0) = r(x)$ and D is a homotopy of the retraction r with the identity and is therefore a deformation retraction. ∎

We note that this lemma actually describes a way to construct a deformation retraction of X onto Y given a deformation and a retraction of X into Y.

Lemma 1.4. *A map $f:X \to Y$ has a left homotopy inverse if and only if X is a retract of M_f.*

Proof. Suppose $r:M_f \to X$ is a retraction. Let $g = r \mid Y$ and $F:X \times I \to X$ be defined by $F(x,t) = r[x,t]$. Then $F(x,0) = r(x) = x$, and $F(x,1) = r[x,1] = r(f(x)) = gf(x)$. Thus F is a homotopy of the identity with gf.

Conversely, suppose that F is a homotopy of the identity map of X with gf. Define $r([x,t]) = F(x,t)$ and $r(y) = g(y)$ for $y \in Y$. Then r is consistently defined, and if $x \in X$, then $x = [x,0] = F(x,0) = r(x)$, and r is a retraction of M_f onto X. ∎

Lemma 1.5. *A map $f:X \to Y$ has a right homotopy inverse if and only if the mapping cylinder M_f deforms into X.*

Proof. Let F be a deformation of M_f into X. Then $g = F \mid Y \times \{1\}$ is a map of Y into X. If $r:M_f \to Y$ is the retraction defined by the map of lemma 1.2, then the map $G = rF \mid Y \times I$ is a homotopy of fg with the identity map 1_Y.

Conversely, let $g:Y \to X$ be a right homotopy inverse to f and let G be a homotopy of 1_Y with fg. Let $H:Y \times I \to M_f$ be defined by $H(y,t) = G(y,2t)$ for $0 \leq t \leq 1/2$ and $H(y,t) = [g(y),2-2t]$ for $1/2 \leq t \leq 1$. Then H is a homotopy of the inclusion $Y \subset M_f$ with the map $g:Y \to X \subset M_f$. To obtain the deformation of M_f into X, we first perform the deformation retraction of M_f into Y given in lemma 1.2, and then apply the homotopy H. ∎

THEOREM 1.6. *A map $f:X \to Y$ is a homotopy equivalence if and only if X is a deformation retract of the mapping cylinder M_f. If D is such a deformation retraction, then $D \mid Y \times \{1\}$ is a homotopy inverse to f and, for any homotopy inverse g, there is a deformation retraction of M_f into X which gives rise to g in this manner.*

Proof. The first statement is spelled out in lemmas 1.4 and 1.5 and the fact that if f has a left and a right homotopy inverse, then either is a two-sided inverse (see 0.3). The second statement follows from the note at the end of lemma 1.3; for if g is homotopy inverse to f and F, G are homotopies, respectively, of fg and gf with the identity, then g and the homotopies F, G define a retraction and a deformation of M_f to X, and the deformation retraction D constructed as in lemma 1.3 has g for its restriction to $Y \times \{1\}$. ∎

THEOREM 1.7. *A map $\boldsymbol{f}:\boldsymbol{X} \to \boldsymbol{Y}$ of n-ads is a homotopy equivalence if and only if its mapping cylinder n-ad $\boldsymbol{M}_f$ has $\boldsymbol{X}$ for a deformation retract.* ∎

2. HOMOTOPY EQUIVALENCE OF ADJUNCTION SPACES

Our first results about homotopy equivalence and type of CW complexes follows from our description of them (II.2.4) as spaces obtained from a discrete set by repeated adjunction of cells. It turns out that if we are given a CW complex and a set of attaching maps, and if we vary each attaching map by a homotopy, the resulting CW complex has the same homotopy type.

We begin by proving a general result about the homotopy type of adjunction spaces.

LEMMA 2.1. *If X is a normal space, if $A \subset X$ is the set of zeros of a continuous map $X \to I$, and if A is a strong deformation retract of a* *neighborhood U of A in X, then $(X\times\{0\})\cup(A\times I)$ is a strong deformation retract of $X\times I$. (Compare with II.7.1.)*

Proof. Let $w:X\to I$ be such that $A=w^{-1}(0)$, and let $h:U\times I\to X$ be a strong deformation of U into A. Assume U is open and let $v:X\to I$ be a Urysohn function such that $v(A)=0$ and $v(X-U)=1$. Then the function $u:X\to I$ defined by $u=\max\{v,w\}$ is an Urysohn function such that $A=u^{-1}(0)$ and $u(X-U)=1$. We will obtain the strong deformation retraction $D:X\times I\times I\to X\times I$ by a modification of the map defined by Young [44]. We thus define:

(i) for $u(x)=0$, $D(x,t,s)=(x,t)$;
(ii) for $0<u(x)\leq 1/2$ and $2u(x)\leq t$, $D(x,t,s)=(h(x,s), t-2su(x))$;
(iii) for $0<u(x)\leq 1/2$ and $t\leq 2u(x)$, $D(x,t,s)=(h(x,st/2u(x)), t(1-s))$;
(iv) for $1/2\leq u(x)\leq 1$, $D(x,t,s)=(h(x,2st(1-u(x))), t(1-s))$; and
(v) for $u(x)=1$, $D(x,t,s)=(x,t(1-s))$.

One easily checks that these maps are consistent, the only difficulty occurring in the matching of formula (i) with formulas (ii) and (iii). We leave the details to the reader. ▮

Corollary 2.2. *If X is a CW complex and $A\subset X$ is a subcomplex, then $(X\times\{0\})\cup(A\times I)$ is a strong deformation retract of $X\times I$.*

Proof. By proposition II.4.3, X is perfectly normal, so that the subcomplex A is the set of zeros of a continuous function $u:X\to I$. By theorem II.6.1, A is a strong deformation of a neighborhood of A. ▮

Theorem 2.3. *Let X be a normal space, and let $A\subset X$ be the set of zeros of a continuous function $X\to I$ and a strong deformation retract of a neighborhood of A. If $h:Y\to Y'$ is a homotopy equivalence, $f:A\to Y$, $f':A\to Y'$, and f' is homotopic to hf, then h extends to a homotopy equivalence $H:Y\cup_f X\to Y'U_{f'}X$. Moreover, each homotopy inverse g of h extends to a homotopy inverse of H.*

Proof. We first reduce the problem to the special case where $f' = hf$. Let $K: A \times I \to Y'$ be a homotopy between f' and hf, and let M_f denote the mapping cylinder of $f: A \to Y$. We extend $h: Y \to Y'$ to a map $h^*: M_f \to Y'$ by defining $h^*([a,t]) = K(a,t)$. Since M_f and Y have the same homotopy type, h^* is a homotopy equivalence. Now define $f^*: A \to M_f$ by $f^*(a) = [a,0]$, and observe that $h^* f^*(a) = h^*([a,0]) = K(a,0) = f'(a)$. Note that there is an obvious strong deformation retraction of $M_f \cup_{f^*} X$ onto $Y \cup_f X$ which is an extension of the strong deformation retraction of M_f onto Y. Thus if we can extend h^* to a homotopy equivalence, we obtain a homotopy equivalence of $Y \cup_f X$ with $Y' \cup_{f'} X$. We will therefore assume that $f' = hf$ in addition to our other hypotheses.

Let $j: X \to Y \cup_f X$ and $j': X \to Y \cup_{f'} X$ be the natural maps which restrict to f and f' on A. We define a map $H: Y \cup_f X \to Y' \cup_f X$ by setting $H(y) = h(y)$ for $y \in Y$, and $Hj(x) = j'(x)$ for $x \in X$. Since $Hj(a) = Hf(a) = hf(a) = f'(a) = j'(a)$, H is consistent; and since it is continuous on both X and Y, it is continuous on $Y \cup_f X$. It remains to prove that H is a homotopy equivalence. We will do this by exhibiting a deformation retraction of the mapping cylinder M_H of H onto $Y \cup_f X$ and applying theorem 1.7.

Let M_h denote the mapping cylinder of $h: Y \to Y'$, and observe that $M_H = M_h \cup_F (X \times I)$, where $F: A \times I \to M_h$ is the composite map

$$A \times I \xrightarrow{f \times 1} Y \times I \subset (Y \times I) \cup Y' \to M_h.$$

The hypotheses on the set $A \subset X$ imply that there is a strong deformation retraction D of $X \times I$ onto $(X \times \{0\}) \cup (A \times I)$ by lemma 2.1. We define a strong deformation retraction of M_H onto $M_h \cup (Y \cup_f X)$ by setting $D'(z,t) = z$ for $z \in M_h$, and $D'(J(x,s),t) = JD((x,s),t)$, where $J: X \times I \to M_h \cup_F (X \times I)$ is the usual map. One easily checks that this is well defined and continuous.

We complete the proof by defining a deformation retraction of $M_h \cup (Y \cup_f X)$ onto $Y \cup_f X$. Since h is a homotopy equivalence, there is a deformation retraction D' of M_h onto Y. Let U be an open neighborhood of A such that there is a strong deformation retraction D of the closure $\bar{U}$ of U onto A in X, and let $u: X \to I$

be a Urysohn function such that $u(A)=1$ and $u(X-U)=0$. Define the deformation retraction D'' of $M_h \cup (Y \cup_f X)$ onto $Y \cup_f X$ as follows: $D''(z,t)=z$ for $0 \leq t \leq 1/2$ and $D''(z,t)=D'(z,2t-1)$ for $1/2 \leq t \leq 1$ whenever $z \in M_h$; $D''(j(x),t)=jD(x,2tu(x))$ for $2tu(x) \leq 1$ and $D''(j(x),t)=D'(jD(x,1),2t-(1/u(x)))$ for $2tu(x) \geq 1$ whenever $x \in \bar{U}$; and $D''(j(x),t)=j(x)$ for $0 \leq t \leq 1$ whenever $x \in X-U$. We leave to the reader the easy proof that D'' is well defined and continuous.

The statement about homotopy inverses follows from theorem 1.7. ∎

COROLLARY 2.4. *Let Y, Y', and X be CW complexes, let $h: Y \rightarrow Y'$ be a homotopy equivalence, let A be a subcomplex of X, and let $f: A \rightarrow Y$ and $f': A \rightarrow Y'$ be maps such that hf is homotopic to f'. Then h extends to a homotopy equivalence of $Y \cup_f X$ with $Y' \cup_{f'} X$.*

COROLLARY 2.5. *Let Y and X be CW complexes, let A be a subcomplex of X, and suppose that $f: A \rightarrow Y$. Then $Y \cup_f X$ has the homotopy type of a CW complex.*

Proof. By theorem II.8.5, the map f is homotopic to a cellular map $f': A \rightarrow Y$. By theorem 2.3 the identity map $Y \rightarrow Y$ extends to a homotopy equivalence $Y \cup_f X \rightarrow Y \cup_{f'} X$. By theorem II.5.11, $Y \cup_{f'} X$ is a CW complex. ∎

COROLLARY 2.6. *Let $h: Y \rightarrow Y'$ be a homotopy equivalence, let $\mathcal{E}^n = \bigcup \{E_\lambda^n \mid \lambda \in \Lambda\}$ be a disjoint union of n-cells, let*

$$\dot{\mathcal{E}}^n = \bigcup \{\dot{E}_\lambda^n \mid \lambda \in \Lambda\} \subset \mathcal{E}^n,$$

let $f: \dot{\mathcal{E}}^n \rightarrow Y$ and $f': \dot{\mathcal{E}}^n \rightarrow Y'$ be maps attaching $\mathcal{E}^n$ *to Y and Y', respectively. If hf is homotopic to f', then h extends to a homotopy equivalence of $Y \cup_f \mathcal{E}^n$ with $Y' \cup_{f'} \mathcal{E}^n$.*

Proof. Identify each n-cell E^n with the unit n-disc D^n. Define $u: \mathcal{E}^n \rightarrow I$ by $u(x)=1-\|x\|$. Then $\dot{\mathcal{E}}^n = u^{-1}(0)$. Also $\dot{\mathcal{E}}^n$ is a strong deformation retract of the neighborhood

$$U = \bigcup \{E_\lambda^n - \{0\} \mid \lambda \in \Lambda\}. \quad \blacksquare$$

THEOREM **2.7.** *Suppose that X and Y are CW complexes satisfying:*

(*1*) $X^0 = Y^0$;

(*2*) *whenever there is a homotopy equivalence* $\varphi^n: X^n \to Y^n$, *and* $f^n: \bigcup_\lambda \dot{E}_\lambda^{n+1} \to X^n$ *and* $g^n: \bigcup_\lambda \dot{E}_\lambda^{n+1} \to Y^n$ *are the maps attaching the* $(n+1)$*-cells, then* $\varphi^n f^n$ *is homotopic to* g^n.

Then there is a homotopy equivalence $\varphi: X \to Y$.

Proof. Define $\varphi^0 =$ identity. Then by induction and corollary 2.6, we obtain homotopy equivalences $\varphi^n: X^n \to Y^n$ for each integer n. The union map $\varphi = \bigcup_n \varphi^n: X \to Y$ is a homotopy equivalence. $\blacksquare$

COROLLARY **2.8.** *Let A be a space which has the homotopy type of a CW complex, and let X be a space which is obtained from A by successive adjunction of cells,* i.e., *there is a sequence* $A = X_{-1} \subset X_0 \subset \cdots \subset X$ *and each* X_k *is obtained from* X_{k-1} *by the adjunction of k-cells. Then X has the homotopy type of a CW complex.* $\blacksquare$

3. WHITEHEAD'S THEOREMS

In this section, we will be concerned with the question: When is a given map between two CW complexes a homotopy equivalence? The answer turns out to be when and only when it is a singular homotopy equivalence, that is, when and only when it induces isomorphisms of the homotopy groups in all dimensions. This case is elementary from our point of view in that the use of the algebraic structure of the homotopy groups is essentially superficial; "isomorphism of homotopy groups" could as well be replaced by "bijection of homotopy sets." We also use the fact that given an inclusion map $i: Y \to X$, the map of homotopy groups is an isomorphism in all dimensions if and only if the relative groups $\pi_k(X,Y)$ all vanish, an immediate consequence of the fact that the homotopy sequence of the pair (X,Y) is exact, or (less efficiently) an easy direct statement to prove without any use of the group structure, using only the criteria given in 0.3 for the relative group vanishing and the definition of the homotopy sets $\pi_k(X)$.

THEOREM 3.1. *Let A be a subcomplex of the CW complex X. If $\pi_k(X,A)=0$ for all k, then A is a strong deformation retract of X.*

Proof. We define $F^0:(X\times\{0\})\cup(A\times I)\to X$ by $F^0(x,0)=x^0$, and $F^0(a,t)=a$ for $a\in A$. We will extend F^0 to a map $F:X\times I\to X$ such that $F(X\times\{1\})\subset A$. The extension will be constructed inductively over the skeletons.

If σ^0 is a vertex in $X-A$, the hypothesis $\pi_0(X,A)=0$ means that there is a path from σ^0 to some point of A. Using such paths we define $F^1:(X\times\{0\})\cup((A\cup X^0)\times I)\to X$. Assume inductively that $F^n:(X\times\{0\})\cup((A\cup X^{n-1})\times I)\to X$ is defined and that F^n extends F. Let σ be an n-cell of X with characteristic map φ. By the induction hypothesis the map $g=F^n((\varphi\mid\dot{E}^n)\times\mathrm{id})$ is a homotopy of $\varphi\mid\dot{E}^n$ with a map $g_1:\dot{E}^n\to A$. Let $G:((E^n\times\{0\})\cup(\dot{E}^n\times I),\dot{E}^n\times\{1\})\to(X,A)$ be the map defined by φ and g. Since the pair $((E^n\times\{0\})\cup(\dot{E}^n\times I),\ \dot{E}^n\times\{1\})$ is homeomorphic to $(E^n,\dot{E}^n)$, the hypothesis $\pi_n(X,A)=0$ implies that there is an extension $G':E^n\times I\to X$ of G such that $G'(E^n\times\{1\})\subset A$. The maps G' and F^n together define an extension

$$F_\sigma^{n+1}:(X\times\{0\})\cup((A\cup X^{n-1}\cup\sigma)\times I)\to X$$

such that $F_\sigma^{n+1}(\sigma\times\{1\})\subset A$. If we let F^{n+1} be the union of all such maps F_σ^{n+1}, then $F^{n+1}:(X\times\{0\})\cup((A\cup X^n)\times I)\to X$ and $F^{n+1}((A\cup X^n)\times\{1\})\subset A$. Finally the union of the maps F^n defines a strong deformation retraction $F:X\times I\to X$ of X onto A. ∎

THEOREM 3.2. *Let $\mathbf{X}$ be a CW n-ad and $\mathbf{A}$ a sub-CW-n-ad. If $\pi_k(X_I,A_I)=0$ for each k and for each corresponding pair of intersections X_I and A_I of the elements of $\mathbf{X}$ and $\mathbf{A}$, then $\mathbf{A}$ is a strong deformation retract of $\mathbf{X}$.*

Proof. As in 3.1, we define $\mathbf{F}^0:(\mathbf{X}\times\{0\})\cup(\mathbf{A}\times I)\to\mathbf{X}$ to be the identity on $\boldsymbol{X}\times\{0\}$ and the projection map (of n-ads) on $\boldsymbol{A}\times I$ onto $\boldsymbol{A}$, and then extend $\boldsymbol{F}^0$ to $\boldsymbol{F}:\boldsymbol{X}\times I\to\boldsymbol{X}$, a map of n-ads, by inductively extending it over the sub-n-ads $(\boldsymbol{X}\times\{0\})\cup((\boldsymbol{A}\cup\boldsymbol{X}^m)\times I)$. This requires slightly more care in the construction of F^m given F^{m-1}. Given F^{m-1}, we first extend it over the *maximal* intersections X_I (the intersections which are nonvoid and contain no other nonvoid intersections). Since $\pi_{m-1}(X_I,A_I)=0$, this is possible, as in the proof of 3.1. Then we extend F^{m-1} over the next largest intersections X_J, etc. After a finite number of steps, we will have extended F^{m-1} over all the proper intersections so that it maps each intersection X_I into the corresponding A_I. Then we extend it over the remaining cells of X^{m-1}. The resulting F^m will map $(X\times\{0\})\cup((A\cup X^{m-1})\times I)$ into X and will send $X^{m-1}\times\{1\}$ into Y, and each $X_I\times\{1\}$ into the corresponding Y_I. It will therefore be a map of n-ads of the required form. ∎

THEOREM 3.3. *Let X and Y have the homotopy types of CW complexes (n-ads), and let f be a map from X to Y. The map f is a homotopy equivalence if and only if it induces isomorphisms of homotopy groups in each dimension (for every intersection X_I to Y_I).*

Proof. If $\phi:X'\to X$ and $\theta:Y\to Y'$ are homotopy equivalences with CW complexes, then $f'=\theta f\phi:X'\to Y'$ also induces isomorphisms of homotopy groups (for every intersection, etc.) in all dimensions. Hence we may assume X and Y are CW complexes to begin with and may replace f by a homotopic cellular map, and assume the mapping cylinder M_f is a CW complex.

Let $p:M_f\to Y$ be the projection map. Because the composition of inclusion of X into M_f with p is f, and because p is a homotopy equivalence, the inclusion map induces isomorphisms in homotopy groups in all dimensions. Thus the relative groups $\pi_k(M_f,X)$ are 0 for all k (and also the relative groups corresponding to the restrictions of to all intersections of the elements of the n-ad $\boldsymbol{X}$). By the previous result, X is a strong deformation retract of M_f, and therefore f is a homotopy equivalence.

The converse is of course trivial. ∎

We recall from III.6.7 that the map $j_X:|S(X)|\to X$ is a

natural transformation that induces homotopy isomorphisms for any space (or n-ad) X.

Corollary 3.4. *A map $f:X\to Y$ induces isomorphisms of homotopy groups in all dimensions (is a singular homotopy equivalence) if and only if and only if $|S(f)|:|S(X)|\to|S(Y)|$ is a homotopy equivalence.*

Proof. From the commutative diagram

$$\begin{array}{ccc} X & \xrightarrow{f} & Y \\ {\scriptstyle j_X}\uparrow & & \uparrow{\scriptstyle j_Y} \\ |S(X)| & \xrightarrow{|S(f)|} & |S(Y)| \end{array}$$

and the fact that f, j_X, and j_Y induce homotopy isomorphisms, we infer that $|S(f)|$ induces homotopy isomorphisms. Since $|S(X)|$ and $|S(Y)|$ are CW complexes, $|S(f)|$ must be a homotopy equivalence.

The converse is trivial. ∎

Corollary 3.5. *The map j_X is a homotopy equivalence if and only if X has the homotopy type of a CW complex.*

Proof. If j_X is a homotopy equivalence, clearly X has the homotopy type of the CW complex $|S(X)|$.

If X has the homotopy type of a CW complex, the fact that j_X induces homotopy isomorphisms together with theorem 3.2 implies that j_X is a homotopy equivalence. ∎

This corollary gives a criterion for a space X to have the homotopy type of a CW complex: j_X must have a homotopy inverse. We shall see below that this can be weakened to require only that j_X have a *left* (one-sided) homotopy inverse.

Corollary 3.6. *A CW complex (or n-ad) has the homotopy type of a CW simplicial complex (n-ad of simplicial complexes).*

Proof. If X is a CW complex, j_X is a homotopy equivalence, and $|S(X)|$ is triangulable. ∎

Definition **3.7.** A space Z *dominates* X if there are maps $f:X \to Z$ and $g:Z \to X$ such that gf is homotopic to the identity map of X. There is an analogous definition for domination of an n-ad by an n-ad.

As an example, a space Z dominates every subspace to which it retracts.

THEOREM **3.8.** *A space (or n-ad) X has the homotopy type of a CW complex if and only if it is dominated by some CW complex (or n-ad).*

Proof. If X has the homotopy type of a CW complex K, let $f:X \to K$ and $g:K \to X$ be a pair of homotopy inverses. Then gf is homotopic to the identity map of X.

Conversely, if X is dominated by a CW complex K, say $f:X \to K$ and $g:K \to X$ have composition homotopic to the identity, then consider the following diagram.

$$\begin{array}{ccccc} X & \xrightarrow{f} & K & \xrightarrow{g} & X \\ {\scriptstyle j}\uparrow\downarrow & & {\scriptstyle j'}\uparrow\downarrow{\scriptstyle k'} & & {\scriptstyle j}\uparrow\downarrow \\ |S(X)| & \xrightarrow{|S(f)|} & |S(K)| & \xrightarrow{|S(g)|} & |S(X)| \end{array}$$

Because K is a CW complex, j' is a homotopy equivalence. Let k' be a homotopy inverse to j', and let $k = |S(g)|k'f$. Note that $|S(g)||S(f)| = |S(gf)| \simeq |S(\mathrm{id})| = \text{identity}$. Thus $jk \simeq gj'k'f \simeq gf \simeq$ identity and $kj \simeq |S(g)||S(f)| \simeq$ identity. Therefore, k is a homotopy inverse to j.

The modification of the argument necessary for the case of n-ads is trivial. ∎

COROLLARY **3.9.** *If the space (or n-ad) X is dominated by (in* *particular, if it is a retract of) a space having the homotopy type of a CW complex (or CW n-ad), then it has the homotopy type of a CW complex (or n-ad).* ∎

4. SIMPLICIAL COMPLEXES WITH THE METRIC TOPOLOGY

In the next sections, in working with simplicial complexes we use the following construction. Let X be a space, let $\mathcal{U} = \{U_\alpha\}$ be an open covering of X, and let $p = \{p_\alpha\}$ be a partition of unity subordinated to $\mathcal{U}$. If Δ is the simplex of $\mathcal{U}$, let

$$N(\mathcal{U}) = \{t_0 U_{\alpha_0} + \cdots + t_n U_{\alpha_n} \in \Delta \mid \textstyle\bigcap_i U_{\alpha_i} \neq \emptyset\}.$$

Then $N(\mathcal{U})$ is a simplicial complex which is a subcell structure of Δ called the *nerve of the covering* $\mathcal{U}$. The partition p determines a canonical map $p^*: X \to N(\mathcal{U})$ according to the formula $p^*(x) = \sum p_\alpha(x) U_\alpha$. For p^* to be continuous, it suffices that the topology be the coarsest in which each barycentric coordinate function be continuous, which turns out to be the standard metric topology on the simplicial complex $N(\mathcal{U})$.

Definition **4.1.** The *coarse topology* on the simplex $\Delta(V)$ is the topology induced by the embedding $b: \Delta(V) \to I^V$ which is given by sending the point $t = \sum t_v v$ into the map $t': V \to I$ for which $t'(v) = t_v$. If K is a simplicial complex with vertex set V, then the coarse topology on K is the one it inherits as a subspace of $\Delta(V)$.

LEMMA **4.2.** *Let K be a simplicial complex with the coarse topology, let $\{b_v\}$ be its family of barycentric coordinates, and let X be a topological space. A function $f: X \to K$ is continuous if and only if the composition $b_v f$ is continuous for each vertex v of K.*

Proof. The map f is continuous if and only if each composition

$$X \xrightarrow{f} K \xrightarrow{\subset} \Delta(V) \xrightarrow{b} I^V \xrightarrow{\pi_v} I$$

is continuous, where V is the vertex set for K and π_v is a projection map. But it is trivial to check that this composition is just $b_v f$. ∎

LEMMA 4.3. *The coarse topology on a simplex (and therefore on each simplicial complex) is metrizable. One metric which induces it is the standard metric* $d(\sum t_v v, \sum t_v' v) = (\sum (t_v - t_v')^2)^{1/2}$.

Proof. A subbase for the coarse topology consists of sets U whose points have vth coordinate in an open subset of I. For a point x of U, pick $\epsilon > 0$ so that the interval $(x_v - \epsilon, x_v + \epsilon) \cap I$ is contained in this open subset of I, where x_v is the vth barycentric coordinate of x. Then the neighborhood $\{y \mid d(y,x) < \epsilon/2\}$ of x lies entirely in U because $d(x,y) < \epsilon/2$ implies $|x_v - y_v| < \epsilon/2$. Thus U is open in the metric topology.

If x is any point, let $N_\epsilon(x)$ be an ϵ-neighborhood of x. If $y \in N_\epsilon(x)$, pick a δ so small that $N_\delta(y) \subset N_\epsilon(x)$. We can choose an r so small that there is a set W of all points such that for a certain n coordinates corresponding to vertices $\{v_1, \cdots, v_n\}$ W consists of all points whose v_i coordinates differ from those of y by no more than r. This will prove that $N_\epsilon(x)$ is a neighborhood of each of its points in the coarse topology, and that the two topologies are identical. To choose r and $v_1, \cdots, v_n$, suppose the nonzero coordinates of y are $y(v_1), \cdots, y(v_n)$. If W is the set of all points whose v_i coordinates differ from those of y be no more than ϵ, then for any $z \in W$, $\sum_{v \neq v_i} z(v) = 1 - \sum z(v_i) < 1 - \sum (y(v_i) - r) = nr$, and therefore $\sum_{v \neq v_i} z(v)^2 < n^2 r^2$. Hence $d(z,y)^2 < nr^2 + n^2 r^2$. Choose $0 < r < \epsilon/(n + n^2)$. ∎

COROLLARY 4.4. *A simplicial complex in the coarse topology is paracompact. The coarse topology is induced on a simplicial complex by embedding into any simplex.* ∎

While there is no general criterion for continuity of a map into a CW complex, we have the following result for canonical maps.

PROPOSITION 4.5. *Let X be a space and let $\mathfrak{U} = \{U_\alpha \mid \alpha \in A\}$ be an* *open cover with* $p = \{p_\alpha \mid \alpha \in A\}$ *a subordinated partition of unity. Also let* p *denote the canonical map* $p: X \to N(\mathfrak{U})$, *where the nerve is given the CW topology. Then* p *is continuous.*

Proof. We will show that for each $x \in X$, there is a neighborhood V_x of x on which p is continuous; this will prove that the map p is continuous on all of X. For the neighborhood V_x, take a neighborhood on which no more than a finite number of functions $p_{\alpha_1}, p_{\alpha_2}, \cdots, p_{\alpha_n}$ are nonzero. Then $p(V_x)$ lies in the subcomplex K of $N(\mathfrak{U})$ with vertices $U_{\alpha_1}, U_{\alpha_2}, \cdots, U_{\alpha_n}$, which is a finite subcomplex of $N(\mathfrak{U})$. On a finite subcomplex of a CW simplicial complex, the CW and coarse metric topologies coincide, so that the map $p \mid V_x$ of V_x into K is continuous if and only if it composes with the barycentric coordinates to give continuous maps. Since for each α_i, $b_{\alpha_i}(p \mid V_x) = p_{\alpha_i}: X \to I$ is continuous, it follows that $p \mid V_x$ is continuous. ∎

Since the barycentric coordinates on a simplicial complex are continuous (they are continuous on each finite dimensional simplex), the CW topology is finer than the coarse (metric) topology. Thus if K is a simplicial complex, let K_{cw} denote K with the CW topology and K_m denote K with the metric topology. If i is the identity map on K, then the map $i: K_{cw} \to K_m$ is continuous. If K is locally finite, it is an easy exercise to check that i^{-1} is also continuous, and therefore that i is a homeomorphism. If K is not locally finite, then i is *not* a homeomorphism, as we see from the following example. Let K consist of the simplicial complex with countably many one dimensional simplexes meeting at a vertex v. Let C be a set consisting of one point from each one simplex, that from the nth having v coordinate $1/n$. In the CW topology, this set is closed (since it meets each cell in a single point, a closed set), whereas in the metric topology v is evidently an accumulation point not in C, and so C is not closed. Note that any simplicial complex not locally finite contains a subcomplex of this sort which is a closed subset and therefore the CW topology must be finer than the metric topology. However, we do obtain the following.

PROPOSITION **4.6.** *Let K be a simplicial complex (or n-ad). Then the identity map $i:K \to K$ is a homotopy equivalence $K_{cw} \to K_m$.*

Proof. To construct a homotopy inverse i', first take a locally finite covering of K by open sets $\{U_v\}$, where U_v contains the vertex v and is contained in the open star of v. For example, $U_v = \{x \mid x(v) > (1/2) \max_w x(w)\}$, for each vertex v. Choose a partition of unity $\{p_v\}$ subordinated to this covering. For example, let

$$p_v(x) = \frac{+\max[(x(v) - (1/2) \max\{x(w)\}), 0]}{\sum \max[(x(v) - (1/2) \max\{x(w)\}), 0]}.$$

The numerator clearly has support contained in U_v, and thus the numerators are a family of functions with locally finite supports. Hence their sum and max, etc., are continuous. Since one of the numerators is nonzero at each point, the sum is nonzero and therefore the quotient is continuous. Let $i'(x)$ be the point of K whose vth barycentric coordinate is $p_v(x)$. The canonical map i' is continuous by 4.5.

For any x, there is a closed simplex containing x and $i'(x)$. Hence the linear homotopy $tx + (1-t)i'(x)$ is a homotopy of $ii' = i'i = i'$ with the identity. This homotopy is continuous in either the metric or the CW topology. Hence i has a two-sided homotopy inverse.

No modification is needed for the case of n-ads since the map i' and the homotopy given above map each closed simplex into itself, and therefore each subcomplex too. ∎

COROLLARY **4.7.** *A CW complex (or n-ad) has the homotopy type of a simplicial complex (or n-ad) with the metric topology.* ∎

5. EQUI-LOCAL CONVEXITY

We will prove that certain function spaces are of CW homotopy type.

Definition **5.1.** A space X is *equi-locally convex* (ELCX) if there is a neighborhood U of the diagonal Δ in $X \times X$, a continuous map $\phi: U \times I \to X$, and an open covering $\{V_\alpha\} = \mathcal{V}$ of X such that:

(i) $\phi(x,x,t) = x$ for all $t \in I$;
(ii) $\phi(x,y,0) = x$, $\phi(x,y,1) = y$;
(iii) $V_\alpha \times V_\alpha \subset U$ for all α; and
(iv) $\phi(V_\alpha \times V_\alpha \times I) \subset V_\alpha$ for each α.

We call a subset W of X convex with respect to ϕ if $W \times W \subset U$ and $\phi(W \times W \times I) \subset W$, and the covering $\mathcal{V}$ is called a convex (open) cover of X.

Lemma **5.2.** (i) *An open subset of a locally convex topological vector space is ELCX.*
(ii) *A simplicial complex in the coarse (metric) topology is ELCX. The covering by open stars is a convex cover.*

Proof. (i) Recall that a topological vector space is locally convex if each point (or just the origin) has a neighborhood base consisting of convex sets, and therefore of open convex sets. If U is an open subset of such a space, let $\{N(x) \mid x \in U\}$ be a covering by open convex neighborhoods contained in U. If $y,x \in N(z)$, let ϕ be the linear path $\phi(x,y,t) = tx + (1-t)y$. With respect to this ϕ, the covering by neighborhoods is a convex cover.

(ii) Let x,y be points of $\mathrm{St}(v_0)$. For each vertex of X, choose the minimum of the coordinates of x and y. Since both x and y have nonzero coordinates corresponding to v_0, at least one of these numbers is nonzero, and since there are only a finite number of coordinates nonzero for either point, the sum of these choices is defined and is positive. Let $s(x,y)$ be the point of X whose v-coordinate is the minimum of the v-coordinates of x and y divided by the sum of all such minima, and let $\phi(x,y,t)$ be the linear path from x to $s(x,y)$ and then to y covering the first half during the time interval $0 \leq t \leq 1/2$, the second during the interval $1/2 \leq t \leq 1$. Since $s(x,y)$ lies in a simplex containing x and also one containing y, this definition makes sense. Finally, since v_0 coordinate of $s(x,y)$ is nonzero, for each point of the path from x to $s(x,y)$

the v_0 coordinate is nonzero. Similarly, the v_0 coordinate of each point of the path from $s(x,y)$ to y is nonzero and therefore the whole path lies in the star of v_0. The continuity of ϕ follows from the fact that the selection of $s(x,y)$ is a continuous function of the points x,y depending as it does on the barycentric coordinates of the points. ∎

THEOREM **5.3.** *If* $\mathbf{X}$ *is an ELCX n-ad with X paracompact, then* $\mathbf{X}$ *has the homotopy type of a CW n-ad.*

Proof. Let $\mathfrak{U}$ be a convex cover of $\mathbf{X}$, and let $\mathfrak{U}^*$ be a star refinement of $\mathfrak{U}$ with the further property that if a set $U^* \in \mathfrak{U}^*$ meets each of the subspaces $X_{i_1}, X_{i_2}, \cdots, X_{i_k}$ of X, then $U^* \cap \bigcap \{X_j \mid j \in \{i_1, i_2, \cdots, i_k\}\} \neq \emptyset$. This is easy to achieve: let $\mathfrak{U}'^*$ be any star refinement, and let $\mathfrak{U}^*$ be the collection of all sets $U'^* - \bigcup \{X_j \mid j \in \{i_1, i_2, \cdots, i_k\}\}$, where $U'^* \cap \bigcap \{X_j \mid j \notin \{i_i, i_2, \cdots, i_k\}\} \neq \emptyset$. Let N_i be the subcomplex of $N(\mathfrak{U})$ with simplexes $\langle U_{i_0}, U_{i_1}, \cdots, U_{i_n}\rangle$, where $X_i \cap \bigcap \{U_j \mid j \in \{i_0, i_i, \cdots, i_n\}\} \neq \emptyset$. Let $\mathbf{N} = (N(\mathfrak{U}^*); N_1, \cdots, N_{n-1})$, and let $p: \mathbf{X} \to \mathbf{N}$ be any canonical map associated with a partition of unity subordinated to $\mathfrak{U}^*$. We will construct a map $q: \mathbf{N} \to \mathbf{X}$ such that the composition qp is homotopic to the identity of $\mathbf{X}$. Then by corollary 3.9, **X** has the homotopy type of a CW n-ad.

Order the sets of $\mathfrak{U}^*$ and from each $U_\alpha^* \in \mathfrak{U}^*$ choose a point u_α^* such that if $U_\alpha^* \cap X_i \neq \emptyset$, then $u_\alpha^* \in X_i$. Define $q: N(\mathfrak{U}) \to X$ as follows. Map the vertex U_α^* of $N(\mathfrak{U}^*)$ into u_α^*. In the simplex $\langle U^*_{\alpha_0}, \cdots, U^*_{\alpha_n}\rangle$ where $U^*_{\alpha_0} < \cdots < U^*_{\alpha_n}$ we define q inductively on $\langle U^*_{\alpha_0}\rangle, \cdots, \langle U^*_{\alpha_0}, \cdots, U^*_{\alpha_k}\rangle$. Suppose that $q \mid \langle U^*_{\alpha_0}, \cdots, U^*_{\alpha_k}\rangle$ is defined and x is a point of the simplex $\langle U^*_{\alpha_0}, \cdots, U^*_{\alpha_{k+1}}\rangle$. Then x has a unique expression $x = tU^*_{\alpha_{k+1}} + (1-t)x_1$, where $x_1 \in \langle U^*_{\alpha_0}, \cdots, U^*_{\alpha_k}\rangle$ and $0 \leq t \leq 1$. Let $q(x) = \phi(u^*_{\alpha_{k+1}}, x_1, t)$. It is easy to check that q is continuous and maps **N** into **X**.

For each $x \in X$, there is a set $U_\alpha \in \mathfrak{U}$ such that $qp(x)$ and x both belong to U_α, since $\mathfrak{U}^*$ is a star refinement of $\mathfrak{U}$. Thus the map defined by $H(x,t) = \phi(x, qp(x), t)$ is a homotopy of qp with the identity of X. ∎

If **X** is a space or a pair, or if the cover $\mathfrak{U}$ satisfies certain special conditions with respect to the subspaces of the n-ad $\boldsymbol{X}$, then we can be more specific.

PROPOSITION **5.4.** (*After Weil*). *Let* $\boldsymbol{X}$ *be a paracompact ELCX* n*-ad with a convex cover* $\mathfrak{U}=\{U_\alpha \mid \alpha \in A\}$ *such that if* $U_\alpha \cap X_j \neq \varnothing$ *for* $j=i_0,i_i, \cdots,i_k$, *then* $U_\alpha \cap \bigcap \{X_j \mid j=i_0,i_1, \cdots,i_k\} \neq \varnothing$. *If the* N_i are *defined as above and* $p:X \to N$ *is a canonical map associated with a partition of unity subordinated to* $\mathfrak{U}$, *then* p *is a homotopy equivalence.*

Proof. Let $N'(\mathfrak{U})$ be the barycentric subdivision of $N(\mathfrak{U})$, which has one vertex $e(I)$ for each nonvoid intersection $\bigcap\{U_\alpha \mid \alpha \in I = \{i_0, \cdots,i_k\}\} = U_I$ and one simplex $\langle e(I_0), \cdots,e(I_k)\rangle$ for each sequence $I_0 \subset \cdots \subset I_k$ such that $U_{I_k} \neq \varnothing$. Define $q(e(I))$ to be any point $u(I) \in U_I$ so that if $U_\beta \cap X_i \neq \varnothing$, then for any I containing β, we have $u(\beta) \in X_i$. Since each U_I is contractible, we can extend q to a map which sends each $\langle e(I_0), \cdots,e(I_k)\rangle$ into U_{I_0}.

For each $x \in X$, let $p(x)$ be in a minimal dimension simplex $\langle e(I_0), \cdots,e(I_k)\rangle$ of $N'(\mathfrak{U})$. Then $qp(x) \in U_{I_0}$, and since $p(x)$ is an interior point, $x \in U_{I_0}$. Thus x and $qp(x)$ always lie in a common set U_α of $\mathfrak{U}$, so that the map defined by $H(x,t) = \phi(x,qp(x),t)$ is a homotopy of qp with the identity of X.

If $x \in N(\mathfrak{U})$ is an interior point of the simplex $\sigma = \langle e(I_0), \cdots,e(I_k)\rangle$ of $N'(\mathfrak{U})$, then $q(x) \in U_{I_0}$. If $y \in U_\alpha$, then $p(y) \in \overline{\mathrm{St}}(U_\alpha)$ in $N(\mathfrak{U})$; thus $x \in \bigcap\{\overline{\mathrm{St}}(U_\alpha) \mid \alpha \in I_0\}$, the union of the closed simplexes of $N(\mathfrak{U})$ having σ as a face. This is a contractible set, and defines an aspherical carrier $\langle e(I_0), \cdots,e(I_k)\rangle \to \bigcap\{\mathrm{St}(U_\alpha) \mid \alpha \in I_0\}$ carrying both pq and the identity map of $N(\mathfrak{U})$. By II.9.2, pq is homotopic to the identity map of $N(\mathfrak{U})$.

COROLLARY **5.5.** *An open subset, or an* n-ad *of open subsets, of a normed linear* (*vector*) *space over the real numbers has the homotopy type of a CW complex, respectively of a CW* n*-ad.*

Proof. A normed linear space has a metrizable topology, hence is paracompact, and is also locally convex and therefore ELCX. Given an n-ad, we can cover the containing subset with (open) convex discs, making sure that we include coverings of the open subsets by (open) convex discs. Then we apply 5.4. ∎

Corollary **5.6.** *A neighborhood retract in a normed linear space, in particular, an absolute neighborhood retract in the category of metric spaces [see Appendix II], has the homotopy type of a CW complex.*

Proof. Corollary 5.5 implies that a retract of an open set in a normed linear space has the homotopy type of a CW complex. A space which is an absolute retract in the category of metric spaces is embeddable as a subset of a normed linear space (Appendix II.9) and is therefore a neighborhood retract in this space. ∎

Corollary **5.7.** *A separable manifold has the homotopy type of a CW complex.*

Proof. In [17] it is proved that a separable manifold is an absolute neighborhood retract. ∎

As an application of theorem 5.4, we prove that certain function spaces have the homotopy type of a CW complex.

Theorem **5.8.** *If* $\mathbf{X} = (X; X_1, \cdots, X_{n-1})$ *has the homotopy type of a CW n-ad and* $\mathbf{C} = (C; C_1, \cdots, C_{n-1})$ *is an n-ad of compact metric spaces, then* $\mathbf{X}^{\mathbf{C}}$*, the space of all continuous functions from* $\mathbf{C}$ *to* $\mathbf{X}$ *with the compact open topology, has the homotopy type of a CW complex.*

Because of theorem 5.4, the theorem follows from the following two lemmas, which prove first that $\mathbf{X}^{\mathbf{C}}$ is paracompact and then that it is ELCX. ∎

Lemma **5.9.** *If* X *is a metrizable space and* C *a compact metric space, then on the function space* X^C*, the compact-open topology is metrizable.*

Proof. This is a standard theorem about function spaces. It is contained for example in [8, p. 270, No. 3]. ∎

LEMMA 5.10. *If $\boldsymbol{X}$ is a metrizable ELCX n-ad and $\boldsymbol{C}$ is compact, then $\boldsymbol{X}^{\boldsymbol{C}}$ with the induced compact-open topology is ELCX.*

Proof. First consider the case of 1-ads, i.e., $\boldsymbol{X}=X$ and $\boldsymbol{C}=C$. Let $\mathcal{V}=\{V_\alpha\}$ be a convex covering of X and ϕ the associated function. Cover X^C by the open sets $[K:V_\alpha]=\{f\in X^C \mid f(K)\subset V_\alpha$ for $K\subset C$ closed and $V_\alpha\in\mathcal{V}\}$. If $f,g\in[K:V_\alpha]$, define $\phi^*(f,g,t)(c)=\phi(f(c),g(c),t)$ for $c\in K$ and $t\in I$. Observe that $\phi^*(f,g,t)\in[K:V_\alpha]$ for all $t\in I$, because for any $c\in K, f(c),g(c)\in V_\alpha$, so that $\phi(f(c),g(c),t)\in V_\alpha$.

In order to see that $\phi^*:[K:V_\alpha]\times[K:V_\alpha]\times I\to[K:V_\alpha]$ is continuous, we recall [Kelley, theorem 5, p. 223] that if C is compact and X^C has the compact open topology, a map $F:Z\to X^C$ is continuous if and only if the map $f':Z\times C\to X$, defined by $f'(z,c)=f(z)(c)$, is continuous. Thus we need to prove the map $\phi^{*\prime}:[K:V_\alpha]\times[K:V_\alpha]\times I\times C\to X$ is a continuous map. The map $\phi^{*\prime}$ can be factored as the following composition

$$[K:V_\alpha]\times[K:V_\alpha]\times C\times I \xrightarrow{k} [K:V_\alpha]\times C\times[K:V_\alpha]\times C\times I$$

$$\xrightarrow{e\times e\times 1} V_\alpha\times V_\alpha\times I \xrightarrow{\phi} V_\alpha\subset X, \quad \text{where} \quad k(f,g,c,t)=(f,c,g,c,t),$$

and e is the (continuous) evaluation map $[K:V_\alpha]\times C\to V_\alpha$. Thus ϕ^* is continuous and X^C is ELCX.

The general case, $n\geq 1$, follows from this special case because $\boldsymbol{X}^{\boldsymbol{C}}=\bigcap X_I{}^{C_I}$. By the special case above, each of the subspaces $X_I{}^{C_I}$ is ELCX, and it is easy to see that their convex structures are the restrictions of those of X^C. The intersection of ELCX subspaces of this type is again an ELCX subspace. ∎

6. *COUNTABLE CW COMPLEXES*

If we restrict our attention to countable CW complexes, we obtain the following result and its corollaries.

THEOREM **6.1.** *For a connected space X, the following are equivalent.*

(i) *The space X has the homotopy type of a countable simplicial complex.*
(ii) *The space X has the homotopy type of a locally finite simplicial complex.*
(iii) *The space X is dominated by a countable CW complex.*
(iv) *The space X has the homotopy type of a CW complex and its homotopy groups are countable.*
(v) *The space X has the homotopy type of a countable CW complex.*

COROLLARY **6.2.** *If X has the homotopy type of a CW complex and is a Lindelöf space, then X has the homotopy type of a countable CW complex.*

Proof. If X is a Lindelöf space, then so is each continuous image of X. If X is dominated by a countable CW complex K, its image in K must be Lindelöf and therefore have a countable carrier. This countable subcomplex of K dominates X. ∎

COROLLARY **6.3.** *An ANR has the homotopy type of a countable CW complex.*

Proof. An ANR is a separable metric space, hence is Lindelöf. By Corollary 5.6, an ANR has the homotopy type of a CW complex. ∎

COROLLARY **6.4.** *If $\mathbf{X}$ is a countable CW n-ad and $\mathbf{C}$ is a compact n-ad, then $\mathbf{X}^{\mathbf{C}}$ has the homotopy type of a countable CW complex.*

Proof. We may assume that X is a countable simplicial n-ad with the coarse topology. Then we know from 5.7 that $\mathbf{X}^{\mathbf{C}}$ has the homotopy type of a CW complex. Since it is also a separable metric space, it is Lindelöf and therefore by 6.2 has the homotopy type of a countable CW complex. ∎

Proof of 6.1. (i) *implies* (ii). Let X be a countable CW simplicial complex. Write X as the union of an increasing sequence of finite subcomplexes X_n. For each n, let the interval $[n, \infty)$ be subdivided into a countable simplicial complex with vertices at the integer points. Let $X' = \mathbf{U}_n X_n \times [n, \infty)$. Then X' is a locally finite simplicial complex. Let $g: X' \to X$ be the obvious projection. We show that g is a homotopy equivalence. By theorem 3.3 it suffices to show that g induces isomorphisms of homotopy groups in all dimensions. If $h: \mathbf{S}^n \to X$ is any map, the image of h lies in some finite subcomplex X_p and therefore h factors through a map h' into $X_p \times [p, \infty)$ composed with g. Therefore, g induces a homotopy surjection. If $f: \mathbf{S}^n \to X'$ projects under g to a null homotopic map, let F be the homotopy of gf to the constant map. The image of f is contained in a finite subcomplex $\mathbf{U}_{i \leq p} X_i \times [i, \infty)$ of X'. Such a finite subcomplex has X_p as a strong deformation retract, and therefore f is homotopic to a map into $X_r \times \{r\}$ for any $r \geq p$. The null homotopy lies in a finite subcomplex X_q of X. If we choose r bigger than either p or q, then f and F both map into X_r and $X_r \times \{r\}$. Hence we can lift F to a null homotopy of f in $X_r \times \{r\}$, and therefore $f \simeq 0$.

(ii) implies (iii) is trivial.

(iii) implies (iv). We observe first from the simplicial approximation theorem that the homotopy groups of a sphere (or any countable simplicial complex) are countable. By successive cellular adjunction it follows that the homotopy groups of any finite CW complex are countable and, therefore, that the groups of a countable CW complex are countable. Then if X is dominated by a countable CW complex, by 3.8 it is of the homotopy type of a CW complex, and its homotopy groups are isomorphic to a direct summand of the countable homotopy groups of the countable complex.

(iv) implies (v). Given any space X, we can construct a CW complex Q and a map $f: Q \to X$ such that f induces homotopy isomorphisms in all dimensions as follows. Choose one vertex for each path component. Call this Q_0, and let $f_0: Q_0 \to X$ map each point into the corresponding path component. Suppose Q_{n-1} constructed and map f_{n-1} such that f_{n-1} induces homotopy isomorphisms in dimensions less than $n-1$ and a surjection in dimension $n-1$, and Q_{n-1} consists of cells of dimension less than n. Let G_{n-1} be the kernel of $(f_{n-1})_\#$, the induced homotopy homomorphism in dimension $n-1$, and let K_n be the cokernel in dimension n. Choose generators $\{\gamma_i\}$ for G_{n-1}, and $\{\kappa_i\}$ for K_n. For each generator γ_i of G_{n-1} adjoin an n-cell by a map in the homotopy class of the generator. For each generator κ of K_n, adjoin an n-sphere to the vertex corresponding to the path component of the class κ. Let f_n be the map f_{n-1} on Q_{n-1}, a map in the class of κ on the sphere adjoined corresponding to κ, and the extension of the null homotopic map on the boundary of the cell adjoined corresponding to γ_i. Then f_n, Q_n satisfy the induction hypotheses. The result, when applied to a space having countable homotopy groups, is a countable CW complex Q and a map $f:Q\to X$ which induces homotopy isomorphisms. If X has the homotopy type of a CW complex, then f is a homotopy equivalence.

(v) implies (i). Proceeding as in the last, we can construct a semisimplicial complex Q and a map $f:Q\to X$ inducing homotopy isomorphisms in each dimension. Given the induction hypothesis that Q_{n-1} is a semisimplicial complex, we make sure at each step that we choose a semisimplicial map of a subdivision of the n-sphere $\mathbf{S}^n$ in the class of each γ_i. Since the adjunction space of a semisimplicial map is a semisimplicial complex, Q_n and f_n will again satisfy the induction hypothesis. The complex Q can be triangulated to a simplicial complex, and if X has countable homotopy groups, then Q and its subdivision will be countable. If X is a CW complex, f will be a homotopy equivalence. ∎

* 7. *FINITE CW COMPLEXES*

It is reasonable to ask whether we can carry theorem 6.1 a step further and obtain a sufficient condition for finite CW homotopy type. It turns out that this is not completely possible, but we will obtain some partial results. We will use some results in homotopy theory, in particular the Hurewicz theorem [35], and some general results on homology that the reader will find in Chapter V. Singular homology groups are assumed to have integral coefficients.

Proposition **7.1.** *For any space X, the following statements are equivalent.*

(i) *The space X has the homotopy type of a finite CW complex.*
(ii) *The space has the homotopy type of a finite simplicial complex.*
(iii) *The space has the homotopy type of the realization of a finite ssc.*

The proof is an easy modification of the material in the proof of theorem 6.1 and will be omitted. ∎

THEOREM 7.2. *For any space X the following are equivalent.*
(i) *The space X has the homotopy type of a* simply connected *finite CW complex.*
(ii) *The space X is dominated by a* simply connected *finite CW complex.*
(iii) *The space X has the homotopy type of a CW complex, is simply connected, and has finitely generated homology.*

Proof. It is trivial that statement (i) implies statement (ii).

The homology groups of a finite CW complex are finitely generated. This follows, for example, from proposition V.2.11, since the singular homology groups can be calculated from the finitely generated cellular chain groups. If X is dominated by a finite CW complex Y, then $H_*(X)$ is a direct summand of $H_*(Y)$ and is therefore finitely generated. Also, $\pi_1(X)$ is isomorphic to a subgroup of $\pi_1(Y)$, which is trivial by hypothesis, so $\pi_1(X)=0$. Thus statement (ii) implies statement (iii).

Finally, we prove that statement (iii) implies statement (i). Let Q_k and $f_k:Q_k\to X$ be as constructed in 6.1 (iv). We may assume $Q_1=x_0$ is a single point since X is simply connected.

We claim that each Q_k is a *finite* CW complex. The complex Q_1 certainly is, and we assume that Q_j is for $1\leq j\leq k$. Replace $f_k:Q_k\to X$ by the inclusion $i_k:Q_k\to M(f_k)$, where $M(f_k)$ is the mapping cylinder of f_k. The homotopy equivalence $p:M(f_k)\to X$ induces homotopy and homology isomorphisms which identify the maps induced by f_k with those induced by i_k. If $i_{k\#}:\pi_k(Q_k)\to \pi_k(M(f_k))$ is the induced map, let $G_k=$ Ker $i_{k\#}$ and $K_k=$ Coker $i_{k\#}$.

By hypothesis, $f_{k\#}$ (and thus $i_{k\#}$) is an isomorphism in homotopy in dimensions less than k, and is an epimorphism in dimension k. Thus $\pi_r(M(f_k), Q_k)=0$ for $r\leq k$. According to the

Hurewicz theorem, the Hurewicz map $\eta:\pi_r(M(f_k),Q_k)\to H_r(M(f_k),Q_k)$ is an isomorphism for $r\leq k+1$. Thus since X and Q_k have finitely generated homology, so does the pair $(M(f_k),Q_k)$, and we conclude that both $\pi_k(M(f_k),Q_k)$ and $\pi_{k+1}(M(f_k),Q_k)$ are finitely generated. Since G_k and K_k are subgroups of $\pi_{k+1}(M(f_k),Q_k)$ and $\pi_k(M(f_k),Q_k)=0$, respectively, we conclude that G_k and K_k are finitely generated. But then Q_{k+1} is obtained from Q_k by the adjunction of finitely many cells and is a finite complex.

Since X has finitely generated homology, there is an n such that $H_k(X)=0$ for $k>n$. Let Q_n and $f_n:Q_n\to X$ be constructed as above with Q_n a finite complex. By using the mapping cylinder $M(f_n)$ and identifying $i_n:Q_n\to M(f_n)$ with f_n, we see that $f_{n*}:H_r(Q_n)\to H_r(X)$ is an isomorphism for $r\leq n-1$ and is an epimorphism for $r=n$.

It follows easily from proposition V.1.8 that $H_n(Q_n)$ is a finitely generated free abelian group. Let $\kappa_1,\cdots,\kappa_p$ be a basis for the kernel of $i_{n*}:H_n(Q_n)\to H_n(M(f_n))$, and let $\delta_1,\cdots,\delta_p\in H_{n+1}(M(f_n),Q_n)$ be such that $\partial_*(\delta_i)=\kappa_i$ for $1\leq i\leq p$. Let $\bar\delta_i=\eta^{-1}(\delta_i)$, where η is the Hurewicz isomorphism, and let $\bar\kappa_i=\partial_\#(\bar\delta_i)$ for $1\leq i\leq p$. Note that $\eta(\bar\kappa_i)=\kappa_i$. Let h_i be a map in the homotopy class $\bar\kappa_i$ for $1\leq i\leq p$, and let $\bar Q=Q_n\cup_{h_1}E^{n+1}\cup_{h_2}\cdots\cup_{h_p}E^{n+1}$. Since i_nh_i is null homotopic, the map i_n extends to $\bar f:\bar Q\to M(f_n)$, and one easily sees that $\bar f_*$ is an isomorphism in dimensions no greater than n, $H_{n+1}(\bar Q)=0$, and thus that $\bar f_*$ is an isomorphism in all dimensions. If we now replace $\bar f$ by an inclusion in a mapping cylinder and apply the Hurewicz theorem, it follows that $\bar f$ induces isomorphisms of homotopy in all dimensions and is thus a homotopy equivalence. ∎

There is no elementary (i.e., geometric as opposed to algebraic) statement such as theorem 7.2 in the general case of finite CW homotopy type, because of complications introduced by the operation of a nontrivial fundamental group. For example, the previous proof would fail because, in applying the Hurewicz theorem, we would have to take account of the operation of $\pi_1(A)$ on $\pi_n(X,A)$.

Wall [37] has studied the question of when a space dominated by a finite CW complex is of the homotopy type of a finite CW complex and has arrived at conclusions which can be generally described as follows. First, the space can be replaced up to homotopy type by a CW complex which has finite skeletons. Such a complex is said to be of *finite type*. Given a CW complex X of finite type which is dominated by a finite CW complex, let π be its fundamental group and Λ be the integral group ring of π. Associated with Λ there is another ring $\tilde{K}_0(\Lambda)$ [33,34] which is determined by the algebraic structure of Λ. If $\tilde{X}$ is the universal covering space of X, which is also a CW complex, then for each k, $H_k(\tilde{X},\tilde{X}^{k-1})$ is a Λ-module, and its Λ-module structure determines an element of $\tilde{K}_0(\Lambda)$ which is an "obstruction" to finiteness. These elements are eventually all zero if and only if X has the homotopy type of a finite CW complex. Moreover, if X is any nonsimply connected finite CW complex, then we can attach cells in high dimensions to get another CW complex which is dominated by a finite CW complex for which this obstruction does not vanish.

An interesting fact illustrating the complexity of this question is proved by Gersten [13]. If X is any CW complex dominated by a finite CW complex, and if A is a finite CW complex with zero Euler characteristic (for example $\mathbf{S}^{2n+1}$), then the product $X \times A$ has the homotopy type of a finite CW complex.

Finally, we remark that in [37], Wall also finds (not generally practical) necessary and sufficient conditions that a CW complex be of the homotopy type of an n-dimensional CW complex ($n \neq 2$). Together with 6.1 they give a sufficient condition that a space have the homotopy type of a finite dimensional CW complex.

CHAPTER V

THE SINGULAR HOMOLOGY OF CW COMPLEXES

Throughout this chapter R will denote a commutative ring with unit $1 \neq 0$.

The purpose of this chapter is to discuss some special properties of the singular homology functor with coefficients in R on the category $\mathcal{W}$ of CW complexes. We are particularly interested in those properties that reflect the geometry of the cell structure.

We assume the reader is familiar with the fact (and proof) that the singular homology functor H_* satisfies the standard six axioms of Eilenberg–Steenrod [10].

We note the continuity property which is valid for singular homology [30] and which is equivalent to the following.

AXIOM C. *If there is a family of subsets* $\{X_\lambda \mid \lambda \in \Lambda\}$ of X *such that*

(i) $X = \bigcup_\lambda X_\lambda$;

(ii) *any compact subset* $K \subset X$ *is contained in a finite union*

$$X_{\lambda_1} \cup X_{\lambda_2} \cup \cdots \cup X_{\lambda_n};$$

then the homology $H_*(X)$ *is the direct limit of the system of* R-*modules* $H_*(X_{\alpha_1} \cup X_{\alpha_2} \cup \cdots \cup X_{\alpha_n})$ *and maps* $i_{\alpha,\beta *}$ *induced by the inclusion maps*

$$i_{\alpha,\beta}: X_{\alpha_1} \cup X_{\alpha_2} \cup \cdots \cup X_{\alpha_n} \to X_{\beta_1} \cup X_{\beta_2} \cup \cdots \cup X_{\beta_m}.$$

We remark that by a modification of the argument in [10] for simplicial complexes, the Eilenberg–Steenrod axioms determine singular homology uniquely on the subcategory $\mathcal{W}_0$ of finite CW complexes. Milnor [30] proved that these, together with axiom C, determine singular homology on the category $\mathcal{W}$ of all CW complexes. Thus, in theory, anything we want to know about the singular homology of CW complexes can be derived from the Eilenberg–Steenrod axioms together with axiom C. As a practical matter, however, we find it convenient to use considerably more than just these axioms. In this chapter we will develop some machinery to perform calculations, a cellular chain and homology theory, and will apply it to several examples. Much of this development parallels lectures given by W. S. Massey. An indication of the alternative procedure, operating directly from the axioms, is given by the treatments of "generalized cohomology theories" as in Dold [5], Dyer [9], Whitehead [38], and others.

We make the following conventions. If $\varnothing = A \subset X$, then we interpret X/A to mean X with an adjoined but disjoint extra point, which we denote by $*$. If we are dealing with spaces with base points, then this added point will be the base point. If X is a CW complex, we require that $*$ be a vertex. Then $X/\varnothing$ is again a CW complex. Note that it is still true that for any map of pairs $f:(X,A) \to (Y,B)$, there is an induced continuous map $\bar{f}: X/A \to Y/B$ such that in the case $A = \varnothing$, $*$ maps into the point $[B]$ of Y/B, and if both A and B are empty, $\bar{f} \mid X = f$ and f maps $*$ to $*$. We also define the cone $c(A)$ to be a single point if $A = \varnothing$, which we may think of as the vertex of the cone. Thus, in every case, whether A is empty or not, we have a collapsing map $p: X \cup c(A) \to X/A$.

1. EXCISION IN THE CW CATEGORY

We first present some material on excision isomorphisms. This section is preparatory for our subsequent calculations.

THEOREM **1.1.** *If A and B are subcomplexes of the CW complex X such that $X = A \cup B$, then the map*

$$e_*: H_*(A, A \cap B) \to H_*(A \cup B, B)$$

induced by inclusion is an isomorphism.

Proof. Let U be a neighborhood of $A \cap B$ in A of which $A \cap B$ is a strong deformation retract (II.6.1). Let $V = B \cup U$; this is a neighborhood of B in X of which B is a strong deformation retract. Consider the commutative diagram

$$\begin{array}{ccccc}
H_*(A,A\cap B) & \xrightarrow{i_{1*}} & H_*(A,A\cap V) & \xrightarrow{=} & H_*(X-(B-(A\cap B)),V-(B-(A\cap B))) \\
\downarrow e_* & & \downarrow & \swarrow e'_* & \\
H_*(A\cup B,B) & \xrightarrow{i_{2*}} & H_*(A\cup B,V) & &
\end{array}$$

where all maps are induced by inclusions. Since $A \cap B$ and B are strong deformation retracts of $V \cap A$ and V, respectively, the maps i_{1*} and i_{2*} are isomorphisms. The map e_*' is an excision isomorphism since $\overline{B-(A\cap B)} \subset B \subset V$. Thus e_* is an isomorphism. ∎

If A and B are subcomplexes of the CW complex X, we will call an inclusion map $e: (A,A\cap B) \to (A\cup B,B)$ an excision map of CW pairs. Note that if A and B are arbitrary subspaces of the CW complex X, then e_* need not be an isomorphism.

Proposition 1.2. *If (X,A) is a cofibration and A is contractible to a point $a_0 \in A$, then the quotient map $p:(X,A)\to(X/A,*)$ is a homotopy equivalence.*

Proof. Let $H': A\times I \to A$ contract A to a_0. Define a map $H'': (X\times\{0\})\cup(A\times I)\to X$ by $H''(a,t) = H'(a,t)$ for $a\in A$ and $H''(x,0) = x$ for $x \in X$. Since (X,A) is a cofibration, H'' extends to $H: X\times I \to X$. Define $h: X/A \to X$ by $hp(x) = H(x,1)$. This makes sense, because if $a \in A$, $H(a,1) = a_0$. Clearly H is a homotopy of hp with the identity map of X. Note that since $H(A\times I) \subset A$, there is an induced map $\bar{H}: (X/A)\times I \to X/A$ such that $\bar{H}(p(x),t) = pH(x,t)$. The map $\bar{H}$ is a homotopy of the identity map of X/A with ph. Both the homotopies H and $\bar{H}$ are homotopies of maps of pairs. ∎

Proposition 1.3. *If (X,A) is a CW pair, there is an isomorphism*

$$\theta: H_*(X,A) \xrightarrow{\approx} \tilde{H}_*(X/A)$$

such that, if $f:(X,A)\to(Y,B)$ *is a map of CW pairs and* $\bar{f}:X/A\to Y/B$ *is the map induced by* f*, then the following diagram is commutative:*

$$\begin{array}{ccc} H_*(X,A) & \xrightarrow{\theta} & \tilde{H}_*(X/A) \\ \downarrow f_* & & \downarrow \bar{f}_* \\ H_*(Y,B) & \xrightarrow{\theta} & \tilde{H}_*(Y/B). \end{array}$$

If $A\neq\varnothing$*, then* θ *is the composition*

$$H_*(X,A)\xrightarrow{p_*}H_*(X/A,*)\approx\tilde{H}(X/A),$$

where $p:(X,A)\to(X/A,*)$ *is the quotient map.*

Proof. When A is empty, we identify $H_*(X,A)$ with $\tilde{H}_*(X/A)$ as follows. Let $\theta':C_*(X)\to C_*(X)\oplus R*$ be defined by $\theta'(x)=x\oplus(-\epsilon(x))*$, where $*$ means the singular 0-simplex of $X/\varnothing$ at $*$, and $R*$ means all R-multiples of this 0-simplex. That this is a natural chain isomorphism is an easy exercise to check.

Let θ be the composition of the isomorphisms

$$H_*(X,A)\xrightarrow{e_*}H_*(X\cup c(A),c(A))\xrightarrow{p'_*}H_*((X\cup c(A))/c(A),*)$$

$$\overset{k}{=}H_*(X/A,*)\approx\tilde{H}_*(X/A),$$

where e is an excision map of CW pairs, and $p':(X\cup c(A),c(A))\to((X\cup c(A))/c(A),*)$ is the quotient map which is a homotopy equivalence by 1.2. Note that if $A\neq\varnothing$, then $kp_*'e_*$ is the map p_*. Commutativity of the diagram follows from naturality of the maps e and p'. ▮

Let $1_n:\Delta^n\to\Delta^n$ be the identity map, and let $j_\#:C_n(\Delta^n)\to C_n(\Delta^n,\dot{\Delta}^n)$ be the quotient map. Clearly $j_\#1_n$ is a cycle in $C_n(\Delta^n,\dot{\Delta}^n)$. Denote the homology class of a cycle z by $\{z\}$.

PROPOSITION **1.4.** *For each* $n \geq 0$,

(a)$_n$ $H_q(\Delta^n, \dot{\Delta}^n) = 0$ *if* $q \neq n$, *and* $H_n(\Delta^n, \dot{\Delta}^n) \approx R$ *is generated by the class* $\{j_\# 1_n\}$;

(b)$_n$ $\tilde{H}_q(\dot{\Delta}^{n+1}) = 0$ *if* $q \neq n$, *and* $\tilde{H}_n(\dot{\Delta}^{n+1}) \approx R$ *is generated by the class* $\{\partial 1_{n+1}\}$;

(c)$_n$ *the boundary operator* $\partial_* : H_{n+1}(\Delta^{n+1}, \dot{\Delta}^{n+1}) \to \tilde{H}_n(\dot{\Delta}^{n+1})$ *maps* $\{j_\# 1_{n+1}\}$ *to* $\{\partial 1_{n+1}\}$.

Proof. The pair $(\Delta^0, \dot{\Delta}^0) = (\Delta^0, \varnothing)$ may be identified with the point space Δ^0. Thus by direct calculation, $H_q(\Delta^0, \dot{\Delta}^0) = 0$ for $q \neq 0$, and $H_0(\Delta^0, \dot{\Delta}^0) \approx R$ is generated by $\{1_0\} = \{j_\# 1_0\}$. This establishes (a)$_0$.

Assume (a)$_n$ and let K^n be the union of all n-faces of $\dot{\Delta}^{n+1}$ except the 0th. Since K^n is contractible, $i_* : \tilde{H}_q(\dot{\Delta}^{n+1}) \to H_q(\dot{\Delta}^{n+1}, K^n)$ is an isomorphism. Let $e_0 : (\Delta^n, \dot{\Delta}^n) \to (\dot{\Delta}^{n+1}, K^n)$ be the linear homeomorphism onto the 0th face of $\dot{\Delta}^{n+1}$. Then e_0 is an excision map of CW pairs, and $e_{0*} : H_q(\Delta^n, \dot{\Delta}^n) \to H_q(\dot{\Delta}^{n+1}, K^n)$ is an isomorphism. Thus $\tilde{H}_q(\dot{\Delta}^{n+1}) \approx H_q(\dot{\Delta}^{n+1}, K^n) \approx H_q(\Delta^n, \dot{\Delta}^n)$, and if $q \neq n$, $\tilde{H}_q(\dot{\Delta}^{n+1}) = 0$, and $\tilde{H}_n(\dot{\Delta}^{n+1}) \approx R$. Moreover, $\partial 1_{n+1} = (-1)^0 e_{0\#} j_\# 1_n$ in $C_n(\dot{\Delta}^{n+1}, K^n)$, and since we are assuming $\{j_\# 1_n\}$ generates $H_n(\Delta^n, \dot{\Delta}^n)$, we see that $\{\partial 1_n\}$ generates $\tilde{H}_n(\dot{\Delta}^{n+1})$. Thus (a)$_n$ implies (b)$_n$.

From the reduced exact homology sequence of the pair $(\Delta^{n+1}, \dot{\Delta}^{n+1})$ and the fact that Δ^{n+1} is contractible, we infer that $\partial_* : H_{q+1}(\Delta^{n+1}, \dot{\Delta}^{n+1}) \to \tilde{H}_q(\dot{\Delta}^{n+1})$ is an isomorphism. Thus if we assume (b)$_n$, $H_{q+1}(\Delta^{n+1}, \dot{\Delta}^{n+1}) = 0$ for $q \neq n$ and $H_{n+1}(\Delta^{n+1}, \dot{\Delta}^{n+1}) \approx R$, from the definition of ∂_*, we see that $\partial_* \{j_\# 1_{n+1}\} = \{\partial 1_{n+1}\}$ which is a generator of $\tilde{H}_n(\dot{\Delta}^{n+1})$. Thus $\{j_\# 1_{n+1}\}$ generates $H_{n+1}(\Delta^{n+1}, \dot{\Delta}^{n+1})$. We have proved that (b)$_n$ implies (a)$_{n+1}$ as well as statement (c)$_n$.

The proposition follows by induction. ∎

For each $n \geq 0$, we choose a fixed homeomorphism of pairs $h_n : (\Delta^n, \dot{\Delta}^n) \to (D^n, \dot{D}^n)$ and define generators $\zeta_n \in H_n(D^n, \dot{D}^n)$ by $\zeta_n = h_{n*}\{j_\# 1_n\}$, and $\xi_n \in \tilde{H}_n(\dot{D}^{n+1})$ by $\xi_n = h_{n+1*}\{\partial 1_{n+1}\}$. Clearly, $\partial_*(\zeta_{n+1}) = \xi_n$, where ∂_* is the boundary operator in the reduced exact homology sequence of the pair $(D^{n+1}, \dot{D}^{n+1})$.

If a CW complex X has a base point $x \in X$, we will always assume that x is a vertex of X. This is not an important restriction by theorem II.6.4.

Definition **1.5.** Let $\{(X_\lambda, x_\lambda) \mid \lambda \in \Lambda\}$ be a family of CW complexes with base point. The *one-point union* or *wedge sum* of this family is the quotient of the disjoint union of the family $\{X_\lambda \mid \lambda \in \Lambda\}$ by the relation identifying all the points x_λ to a single point x^*. Clearly the one-point of CW complexes is a CW complex. We will denote the one-point union by $\bigvee\{(X_\lambda, x_\lambda) \mid \lambda \in \Lambda\}$ or by $\bigvee_{\lambda \in \Lambda} X_\lambda$. For each $\lambda \in \Lambda$ there is an inclusion map

$$(X_\lambda, x_\lambda) \xrightarrow{i_\lambda} (\bigvee_{\lambda \in \Lambda} X_\lambda, x^*)$$

and a projection map

$$(\bigvee_{\lambda \in \Lambda} X_\lambda, x^*) \xrightarrow{j_\lambda} (X_\lambda, x_\lambda)$$

such that $j_\lambda i_\mu =$ constant at x_λ for $\mu \neq \lambda$, and $j_\lambda i_\lambda =$ identity X_λ.

Lemma **1.6.** *Let $M_i, 0 \leq i \leq 4$ be R-modules, and let i_1, i_2, j_1, j_2, k_1, k_2 be R-linear maps such that in the diagram*

$$\begin{array}{ccccc} & & M_1 & & \\ & & \downarrow{\scriptstyle i_1} & \searrow{\scriptstyle k_1} & \\ M_2 & \xrightarrow{i_2} & M_0 & \xrightarrow{j_2} & M_4 \\ & \searrow{\scriptstyle k_2} & \downarrow{\scriptstyle j_1} & & \\ & & M_3 & & \end{array}$$

the row and column are exact and $j_2 i_1 = k_1$ and $j_1 i_2 = k_2$ are isomorphisms. Then i_1 and i_2 are injections, and $M_0 = i_1(M_1) \oplus i_2(M_2)$. The maps j_1 and j_2 are projections on the direct summands.

The proof is left to the reader. ∎

Proposition **1.7.** *Let* $\{(X_\lambda, x_\lambda) \mid \lambda \in \Lambda\}$ *be a family of CW com-*

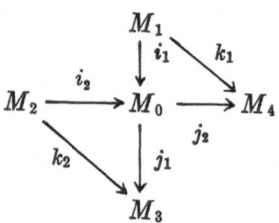

plexes with base point. For each n, the maps $i_\lambda : X_\lambda \to X = \bigvee_{\lambda \in \Lambda} X_\lambda$ induce an R-linear isomorphism

$$I_\Lambda : \sum \oplus \{\tilde{H}_n(X_\lambda) \mid \lambda \in \Lambda\} \to \tilde{H}_n(X).$$

The collapsing map $j_\mu : X \to X_\mu$ induces the projection of the direct sum $\tilde{H}_n(X)$ onto the summand $\tilde{H}_n(X_\mu)$.

Proof. First consider the one-point union $X_1 \vee X_2$ of two CW complexes. Since $(X_1 \vee X_2)/X_1 = X_2$ and $(X_1 \vee X_2)/X_2 = X_1$, we have a commutative diagram

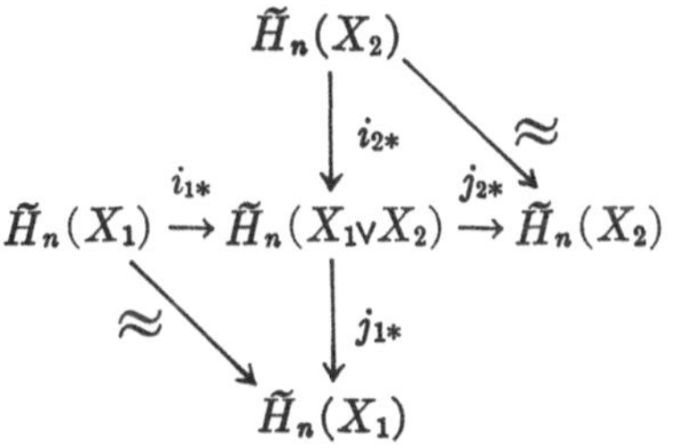

in which each row and column are exact. By lemma 1.6,

$$\tilde{H}_n(X_1 \vee X_2) = i_{1*}(\tilde{H}_n(X_1)) \oplus i_{2*}(\tilde{H}_n(X_2)).$$

An easy induction establishes 1.7 in the case that Λ is a finite set.

In the case Λ is infinite, let $\mathfrak{F}$ be the collection of all $X_\Delta = \bigvee_{\lambda \in \Delta} X_\lambda$, where Δ is a finite subset of Λ. The family $\mathfrak{F}$ satisfies the conditions of axiom C, so that $\tilde{H}_n(X)$ is the direct limit of the system of modules $\tilde{H}_n(X_\Delta)$ and linear maps induced by the inclusions $i_\Delta : X_\Delta \to X$. But we have an isomorphism $I_\Delta : \sum \oplus \{\tilde{H}_n(X_\lambda) \mid \lambda \in \Delta\} \to \tilde{H}_n(X_\Delta)$ for each finite Δ and an easy calculation shows that the direct limit of such a system is the direct sum $\sum \oplus \{\tilde{H}_n(X_\lambda) \mid \lambda \in \Lambda\}$, and the isomorphisms I_Δ pass to an isomorphism $I_\Lambda : \sum \oplus \{\tilde{H}_n(X_\lambda) \mid \lambda \in \Lambda\} \to \tilde{H}_n(X)$. ∎

Since the maps $i_\lambda : X_\lambda \to X = \bigvee_{\lambda \in \Lambda} X_\lambda$ induce the injective direct sum decomposition of 1.7, we may write $\tilde{H}_n(X)$ as an internal direct sum $\tilde{H}_n(X) = \sum \oplus \{i_{\lambda *}(\tilde{H}_n(X_\lambda)) \mid \lambda \in \Lambda\}$.

We remark that proposition 1.7 is not valid for arbitrary spaces because we used the excision property for CW pairs (theorem 1.1). In general, in order that proposition 1.7 hold, we need a condition on the base point $x_\lambda \in X_\lambda$. For example, if the base point x_λ has a neighborhood of which it is a strong deformation retract, the proposition is valid.

We now apply this result to a specific calculation.

PROPOSITION 1.8. *Let X be a CW complex with n-cells $\{\sigma_\lambda \mid \lambda \in \Lambda\}$, and let $\varphi_\lambda : (D^n, \dot{D}^n) \to (X^n, X^{n-1})$ be a characteristic map for σ_λ. Then*

(i) $H_q(X^n, X^{n-1}) = 0$ *for* $q \neq n$;

(ii) $H_n(X^n, X^{n-1}) = \sum \oplus \{\varphi_{\lambda *}(H_n(D^n, \dot{D}^n)) \mid \lambda \in \Lambda\}$ *is a free R-module with basis* $\{\varphi_{\lambda *}(\zeta_n) \mid \lambda \in \Lambda\}$;

(iii) $H_q(X^n) = 0$ *for* $q > n$;

(iv) $H_q(X, X^n) = 0$ *for* $q \leq n$.

Proof. For each $\lambda \in \Lambda$ let $\varphi_\lambda' : (D^n, \dot{D}^n) \to (\sigma_\lambda, \dot{\sigma}_\lambda)$, and let $i_\lambda : (\sigma_\lambda, \dot{\sigma}_\lambda) \to (X^n, X^{n-1})$ be the inclusion map so that $\varphi_\lambda = i_\lambda \varphi_\lambda'$. Then there is an induced homeomorphism $\bar{\varphi}_\lambda' : D^n/\dot{D}^n \to \sigma_\lambda/\dot{\sigma}_\lambda$ and induced inclusions $\bar{\imath}_\lambda : \sigma_\lambda/\dot{\sigma}_\lambda \to X^n/X^{n-1} = \bigvee_{\lambda \in \Lambda} (\sigma_\lambda/\dot{\sigma}_\lambda)$. From the commutative diagram

$$\begin{array}{ccccccc}
H_q(D^n, \dot{D}^n) & \xrightarrow{\varphi_{\lambda *}'} & H_q(\sigma_\lambda, \dot{\sigma}_\lambda) & \xrightarrow{i_{\lambda *}} & H_q(X^n, X^{n-1}) & \xrightarrow{j_{\mu *}} & H_q(X^n, X^n - (\sigma_\mu - \dot{\sigma}_\mu)) \\
\theta & & \theta & & \theta & & \theta \\
\tilde{H}_q(D^n/\dot{D}^n) & \xrightarrow{\bar{\varphi}_{\lambda *}'} & \tilde{H}_q(\sigma_\lambda/\dot{\sigma}_\lambda) & \xrightarrow{\bar{\imath}_{\lambda *}} & \tilde{H}_q(X^n/X^{n-1}) & \xrightarrow{\bar{\jmath}_{\mu *}} & \tilde{H}_q(\sigma_\mu/\dot{\sigma}_\mu)
\end{array}$$

and 1.7 we conclude that

$$\tilde{H}_q(X^n/X^{n-1}) = \sum \oplus \{\bar{\varphi}_{\lambda *}(\tilde{H}_q(D^n/\dot{D}^n)) \mid \lambda \in \Lambda\}.$$

By 1.3, the maps θ are isomorphisms, so that $H_q(X^n, X^{n-1}) = \sum \oplus \{\varphi_{\lambda *}(H_q(D^n, \dot{D}^n)) \mid \lambda \in \Lambda\}$. Statements (i) and (ii) follow from 1.4. Note that the maps $\bar{\jmath}_{\mu *} : \tilde{H}_q(X^n/X^{n-1}) \to \tilde{H}_q(\sigma_\mu/\dot{\sigma}_\mu)$ provide a projective direct sum decomposition of $\tilde{H}_q(X^n/X^{n-1})$.

To prove (iii), consider a portion of the exact sequence of the pair (X^n,X^{n-1}):

$$H_{q+1}(X^n,X^{n-1}) \to H_q(X^{n-1}) \xrightarrow{i_*} H_q(X^n) \to H_q(X^n,X^{n-1}).$$

Part (i) implies that i_* is an isomorphism for $q>n$. By induction we establish that $H_q(X^{n-k}) \xrightarrow{i'_*} H_q(X^n)$ is an isomorphism for $q>n$. Taking $k=n+1$, we see $0=H_q(X^{-1})\approx H_q(X^n)$ for $q>n$.

Part (iv) follows when we first establish inductively that $H_q(X^{n+k},X^n)\approx \tilde{H}_q(X^{n+k}/X^n)=0$ for $q\leq n$, and then observe that $H_q(X,X^n)\approx \tilde{H}_q(X/X^n)$ is the direct limit of the modules $\tilde{H}_q(X^{n+k}/X^n)$ by axiom C. ∎

We mention the following immediate corollary to part (iii).

Corollary 1.9. *If X is a CW complex of dimension n, then $H_q(X)=0$ for $q>n$.* ∎

2. CELLULAR HOMOLOGY

In this section we define a homology theory that reflects the cellular structure of a CW complex X.

Definition **2.1.** Let X be a CW complex with cells $\mathcal{S}$. The *cellular n-chain module* of $(X,\mathcal{S})$ is the module $C_n(\mathcal{S})=H_n(X^n,X^{n-1})$. By proposition 1.8 (ii), $C_n(\mathcal{S})$ is a free R-module. For each integer n we define an R-linear map $\partial_n:C_n(\mathcal{S})\to C_{n-1}(\mathcal{S})$ as the composite

$$C_n(\mathcal{S})=H_n(X^n,X^{n-1}) \xrightarrow{\partial_*} H_{n-1}(X^{n-1}) \xrightarrow{j_*} H_{n-1}(X^{n-1},X^{n-2})=C_{n-1}(\mathcal{S}),$$

i.e., ∂_n is the boundary operator of the exact sequence of the triple (X^n,X^{n-1},X^{n-2}).

Proposition **2.2.** *If $(X,\mathcal{S})$ is a CW complex, the family $C_*(\mathcal{S})=\{C_n(\mathcal{S}),\partial_n\}$ is a free chain complex over R.*

Proof. We need only prove that $\partial_{n-1}\partial_n = 0$. Note that the composite $\partial_{n-1}\partial_n$ may be factored as

$$H_n(X^n,X^{n-1}) \xrightarrow{\partial_*} H_{n-1}(X^{n-1}) \xrightarrow{j_*} H_{n-1}(X^{n-1},X^{n-2}) \xrightarrow{\partial'_*} H_{n-2}(X^{n-2})$$
$$\xrightarrow{j'_*} H_{n-2}(X^{n-2},X^{n-3}),$$

where the maps ∂'_* and j_* are adjacent maps in the exact sequence of the pair (X^{n-1},X^{n-2}). Consequently, $\partial'_* j_* = 0$, which implies $\partial_{n-1}\partial_n = 0$. ∎

The homology modules of this chain complex will be denoted by $H_n(\mathcal{S})$. The graded module $H_*(\mathcal{S}) = \{H_n(\mathcal{S})\}$ is the *cellular homology* of the CW complex $(X,\mathcal{S})$.

Let $(X,\mathcal{S})$ and $(Y,\mathcal{S}')$ be CW complexes, and let $f:X\to Y$ be a cellular map, i.e., $f(X^n) \subset Y^n$. Then f induces an R-linear map

$$f_\#:C_n(\mathcal{S}) = H_n(X^n,X^{n-1}) \xrightarrow{f_*} H_n(Y^n,Y^{n-1}) = C_n(\mathcal{S}').$$

PROPOSITION **2.3.** *If* $(X,\mathcal{S})$ *and* $(Y,\mathcal{S}')$ *are CW complexes and* $f:X\to Y$ *is a cellular map, the induced map* $f_\#:C_n(\mathcal{S})\to C_n(\mathcal{S}')$ *is a chain map, i.e.,* $\partial'_n f_\# = f_\#\partial_n$.

Proof. This follows immediately from the naturality of the boundary operator of the exact sequence of a triple with respect to maps of triples. ∎

A standard result is the following corollary.

COROLLARY **2.4.** *If* $(X,\mathcal{S})$ *and* $(Y,\mathcal{S}')$ *are CW complexes and* $f:X\to Y$ *is a cellular map,* f *induces an* R*-linear map* $f_*:H_n(\mathcal{S})\to H_n(\mathcal{S}')$ *for each integer* n. ∎

Of particular interest is the case of a subcomplex $(A,\mathfrak{J})$ of a CW complex $(X,\mathcal{S})$. In this case, the inclusion map $i:A\to X$ is a cellular map.

Proposition 2.5. *If* $(A,\mathfrak{I})$ *is a subcomplex of the CW complex* $(X,\mathcal{S})$, *then*

$$0 \to H_n(A^n,A^{n-1}) \xrightarrow{i_*} H_n(X^n,X^{n-1}) \xrightarrow{j_*} H_n(X^n \cup A, X^{n-1} \cup A) \to 0$$

is a short exact sequence of free R-modules, hence is a split sequence.

Proof. Consider the commutative diagram

$$\begin{array}{ccccc} H_n(A^n,A^{n-1}) & \xrightarrow{i_*} & H_n(X^n,X^{n-1}) & \xrightarrow{j_*} & H_n(X^n\cup A,X^{n-1}\cup A) \\ \approx \downarrow e_* & & \downarrow = & & \approx \downarrow e'_* \\ H_n(X^{n-1}\cup A^n,X^{n-1}) & \xrightarrow{i'_*} & H_n(X^n,X^{n-1}) & \xrightarrow{j'_*} & H_n(X^n,X^{n-1}\cup A^n). \end{array}$$

The lower sequence is part of the exact sequence of a triple and the maps e_* and e'_* are excision isomorphisms. By proposition 1.7 (i), $0 = H_{n-1}(A^n,A^{n-1}) \approx H_{n-1}(X^{n-1}\cup A^n,X^{n-1})$, and j'_* is an epimorphism. Since $\theta: H_{n+1}(X^n,X^{n-1}\cup A^n) \to \tilde{H}_{n+1}(X^n/(X^{n-1}\cup A^n))$ is an isomorphism and $X^n/(X^{n-1}\cup A^n)$ is an n-dimensional complex, we conclude that $H_{n+1}(X^n,X^{n-1}\cup A) = 0$. Thus i'_* is a monomorphism. Now note that for each pair (Y,Z) appearing in the diagram, Y/Z is a one-point union of spheres, so that $H_*(Y,Z)$ is a free module. It is clear that the isomorphisms e_* and e'_* identify the upper row with the lower so that the former is exact and splits. ∎

Observe that $(X/A)^n = (X^n \cup A)/A$, so $(X/A)^n/(X/A)^{n-1}$ is homeomorphic to $(X^n\cup A)/(X^{n-1}\cup A)$. Therefore we have $H_n(X^n\cup A,X^{n-1}\cup A) \approx H_n((X/A)^n,(X/A)^{n-1})$.

If we define $C_n(\mathcal{S},\mathfrak{I}) = H_n(X^n\cup A,X^{n-1}\cup A)$, the sequence

$$0 \to C_n(\mathfrak{I}) \xrightarrow{i_\#} C_n(\mathcal{S}) \xrightarrow{j_\#} C_n(\mathcal{S},\mathfrak{I}) \to 0$$

is exact. We define $\partial'_n: C_n(\mathcal{S},\mathfrak{I}) \to C_{n-1}(\mathcal{S},\mathfrak{I})$ as the composition

$$H_n(X^n\cup A,X^{n-1}\cup A) \to H_{n-1}(X^{n-1}\cup A)$$
$$\to H_{n-1}(X^{n-1}\cup A,X^{n-2}\cup A),$$

i.e., as the boundary operator of the triple

$$(X^n \cup A, X^{n-1} \cup A, X^{n-2} \cup A).$$

Standard arguments yield

LEMMA 2.6. *The maps* ∂'_n *have the properties*
(i) $\partial'_{n-1}\partial'_n = 0$;
(ii) $j_\#$ is a chain map. ∎

Applying this lemma in the usual way we obtain the following.

PROPOSITION 2.7. *If* $(A, \mathfrak{I})$ *is a subcomplex of the CW complex* $(X, \mathcal{S})$, *there is an exact sequence*

$$\cdots \to H_n(\mathfrak{I}) \xrightarrow{i_*} H_n(\mathcal{S}) \xrightarrow{j_*} H_n(\mathcal{S}, \mathfrak{I}) \xrightarrow{\partial_*} H_{n-1}(\mathfrak{I}) \to \cdots,$$

and the map ∂_* *is natural with respect to maps induced by cellular maps of CW pairs.* ∎

PROPOSITION 2.8. *If* $(X, \mathcal{S})$ *and* $(Y, \mathcal{S}')$ *are CW complexes and* $f_0, f_1 : X \to Y$ *are homotopic cellular maps, then* $f_{0*} = f_{1*} : H_n(\mathcal{S}) \to H_n(\mathcal{S}')$.

Proof. By corollary II.8.8, f_0 and f_1 are homotopic by a cellular homotopy, and $f_{0*} = f_{1*} : H_n(X^n, X^{n-1}) \to H_n(Y^n, Y^{n-1})$. Thus $f_{0\#} = f_{1\#} : C_n(\mathcal{S}) \to C_n(\mathcal{S}')$, and the result follows. ∎

For completeness we state the following which follows immediately by an argument similar to the argument in the proof of theorem 1.1.

PROPOSITION 2.9. *Let* $(X, \mathcal{S})$ *be a CW complex and let* $(A, \mathfrak{I})$ *and* $(B, \mathfrak{I}')$ *be subcomplexes such that* $(A \cup B, \mathfrak{I} \cup \mathfrak{I}') = (X, \mathcal{S})$. *Then the excision* $e:(B,A\cap B)\to(X,A)$ *induces isomorphisms*

$$e_*:H_n(\mathcal{T}',\mathcal{T}\cap\mathcal{T}')\to H_n(\mathcal{S},\mathcal{T})$$

for all n. ∎

Combining the propositions of this section we finally obtain the following.

THEOREM **2.10.** *Cellular homology is a homology functor from the category of CW complexes and cellular maps to the category of R-modules and R-linear maps.* ∎

The main result of this section is the relation of the cellular homology groups, which are combinatorial in nature, with the topologically invariant singular homology groups.

PROPOSITION **2.11.** *If* $(A,\mathcal{T})$ *is a subcomplex of the CW complex* $(X,\mathcal{S})$, *then for each integer* n *there is an isomorphism*

$$\Psi_n:H_n(\mathcal{S},\mathcal{T})\to H_n(X,A)$$

which is natural with respect to cellular maps of pairs of CW complexes.

Proof. In order to shorten the notation, let $\bar{A}^n=X^n\cup A$. From commutativity of the diagram

$$\begin{array}{ccccc}
 & & H_{n-1}(\bar{A}^{n-1}) & & \\
 & \overset{\partial'_{n*}}{\nearrow} & & \overset{j'_{n-1*}}{\searrow} & \\
C_n(\mathcal{S},\mathcal{T})=H_n(\bar{A}^n,\bar{A}^{n-1}) & & \Big\downarrow & & H_{n-1}(\bar{A}^{n-1},\bar{A}^{n-2})=C_{n-1}(\mathcal{S},\mathcal{T}). \\
 & \underset{\partial_{n*}}{\searrow} & & \underset{j_{n-1*}}{\nearrow} & \\
 & & H_{n-1}(\bar{A}^{n-1},A). & &
\end{array}$$

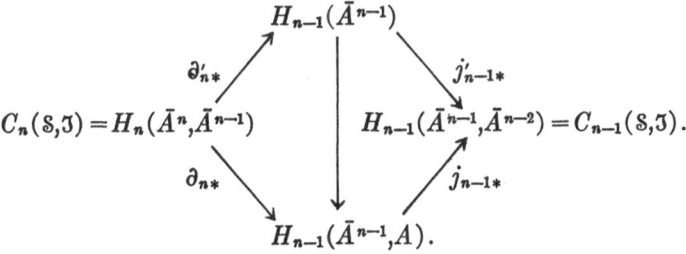

We see that either the upper or the lower composition will serve as the boundary operator for $C_*(\mathcal{S},\mathfrak{J})$. From the exact sequence of the triple $(\bar{A}^{n-1},\bar{A}^{n-2},A)$, the fact that $\bar{A}^{n-2}/A = X^{n-2}/A^{n-2}$ is a complex of dimension less than $n-1$, the fact that

$$H_{n-1}(\bar{A}^{n-2},A) \approx \tilde{H}_{n-1}(\bar{A}^{n-2}/A),$$

and corollary 1.9, we conclude that j_{n-1*} is a monomorphism. We now calculate the image $B_n(\mathcal{S},\mathfrak{J})$ of the boundary operator $j_{n*}\partial_{n+1*}$ and the kernel $Z_n(\mathcal{S},\mathfrak{J})$ of the boundary operator $j_{n-1*}\partial_{n*}$. Since for each n, j_{n*} is a monomorphism, $B_n(\mathcal{S},\mathfrak{J}) = j_{n*}(\mathrm{Im}\ \partial_{n+1*})$; and $Z_n(\mathcal{S},\mathfrak{J}) = \mathrm{Ker}\ j_{n-1*}\partial_{n*} = \mathrm{Ker}\ \partial_{n*} = \mathrm{Im}\ j_{n*}$. These expressions for the cycles and bounding cycles of $C_n(\mathcal{S},\mathfrak{J})$ are natural with respect to maps induced on the chains of cellular maps.

We define $\psi_n: Z_n(\mathcal{S},\mathfrak{J}) \to H_n(X,A)$ to be the map $i_{n*}(j_{n*})^{-1}$, where $i_{n*}: H_n(\bar{A}^n,A) \to H_n(X,A)$ is induced by the inclusion map $i_n: (\bar{A}^n, A) \to (X, A)$. We first prove that ψ_n is an epimorphism, and then prove that $\mathrm{Ker}\ \psi_n = B_n(\mathcal{S},\mathfrak{J})$. This implies that ψ_n induces an isomorphism $\Psi_n: H_n(\mathcal{S},\mathfrak{J}) \to H_n(X,A)$. It is clear that this construction is natural with respect to cellular maps of CW pairs.

From proposition 1.8 (iv), $H_q(X,\bar{A}^{n+1}) = 0$ for $q \leq n+1$ and $H_n(\bar{A}^{n+1},\bar{A}^n) = 0$. Thus if we factor i_n as the composite

$$(\bar{A}^n,A) \xrightarrow{i'_n} (\bar{A}^{n+1},A) \xrightarrow{i_{n+1}} (X,A),$$

the fact that $i'_{n*}: H_n(\bar{A}^n,A) \to H_n(\bar{A}^{n+1},A)$ is an epimorphism, and $i_{n+1*}: H_n(\bar{A}^{n+1},A) \to H_n(X,A)$ is an isomorphism follow from the exact sequences of the triples $(\bar{A}^{n+1},\bar{A}^n,A)$ and $(X,\bar{A}^{n+1},A)$, respectively. But then $\psi_n: Z_n(\mathcal{S},\mathfrak{J}) \to H_n(X,A)$ is an epimorphism. Since i_{n+1*} is an isomorphism in dimension n, $\mathrm{Ker}\ \psi_n = \mathrm{Ker}\ i'_{n*}$. From the exact sequence of the triple $(\bar{A}^{n+1},\bar{A}^n,A)$, we see that in dimension n $\mathrm{Ker}\ i'_{n*} = \mathrm{Im}\ \partial_{n+1*} = j_{n*}^{-1}(B_n(\mathcal{S},\mathfrak{J}))$. Thus $\mathrm{Ker}\ \psi_n = B_n(\mathcal{S},\mathfrak{J})$. ∎

PROPOSITION **2.12.** *Let* $(A,\mathfrak{J})$ *be a subcomplex of the CW complex*

$(X,\mathcal{S})$. *Then for each* n *the diagram*

$$\begin{array}{ccc} H_{n+1}(\mathcal{S},\mathcal{T}) & \xrightarrow{\partial_*} & H_n(\mathcal{T}) \\ \big\downarrow{\scriptstyle \Psi_{n+1}} & & \big\downarrow{\scriptstyle \Psi_n} \\ H_{n+1}(X,A) & \xrightarrow{\bar{\partial}_*} & H_n(A) \end{array}$$

is commutative.

Proof. Consider the commutative diagram

$$\begin{array}{ccccccc}
H_{n+1}(X,A) & \longleftarrow & \longleftarrow & \xleftarrow{\;i'_*\;} & \longleftarrow & \longleftarrow & H_{n+1}(\bar{A}^{n+1},A) \\
\big\downarrow & & & & & & \big\downarrow{\scriptstyle k_*} \\
\big\downarrow{\scriptstyle \bar{\partial}_*} & & H_{n+1}(A^{n+1},A^n) & \xrightarrow{\alpha} & H_{n+1}(X^{n+1},X^n) & \xrightarrow{\beta} & H_{n+1}(\bar{A}^{n+1},\bar{A}^n) \\
\big\downarrow & & \big\downarrow & & \big\downarrow{\scriptstyle \partial'_*} & & \big\downarrow{\scriptstyle \partial''_*} \\
H_n(A) & \xleftarrow{i_*} & H_n(A^n) & \xrightarrow{\alpha'} & H_n(X^n) & \xrightarrow{\beta'} & H_n(\bar{A}^n) \\
& \searrow & & \xrightarrow{\qquad \eta \qquad} & & \nearrow & \big\downarrow \\
& & \big\downarrow{\scriptstyle j_*} & & \big\downarrow{\scriptstyle j'_*} & & \big\downarrow \\
& & H_n(A^n,A^{n-1}) & \xrightarrow{\alpha''} & H_n(X^n,X^{n-1}) & \xrightarrow{\beta''} & H_n(\bar{A}^n,\bar{A}^{n-1}),
\end{array}$$

with the additional commutativity relations $\eta i_* = \beta'\alpha'$ and $\eta\bar{\partial}_* i'_* = \partial''_* k_*$. All maps are boundaries or are induced by inclusion.

If $z \in H_{n+1}(\mathcal{S},\mathcal{T})$ is represented by $k_*(z) \in H_{n+1}(\bar{A}^{n+1},\bar{A}^n)$, then $\partial_*(z)$ is represented by $x \in H_n(A^n,A^{n-1}) = C_n(\mathcal{T})$, where $\alpha''(x) = j'_*\partial'_*(y)$ and $\beta(y) = k_*(z)$. Since x is a cycle, $x = j_*(w)$ for $w \in H_n(A^n)$, and since j'_* is a monomorphism, $\alpha'(w) = \partial'_*(y)$. By definition of Ψ_n, $\Psi_n\partial_*(z) = i_*(w)$. Note that $\bar{\partial}_*\Psi_{n+1}(z) = \bar{\partial}_* i'_*(z)$. Using the additional commutativity relations stated,

$$\eta\Psi_n\partial_*(z) = \eta i_*(w) = \beta'\alpha'(w) = \beta'\partial'_*(y) = \partial''_*\beta(y)$$

$$= \partial''_* k_*(z) = \eta\bar{\partial}_* i'_*(z) = \eta\bar{\partial}_*\Psi_{n+1}(z).$$

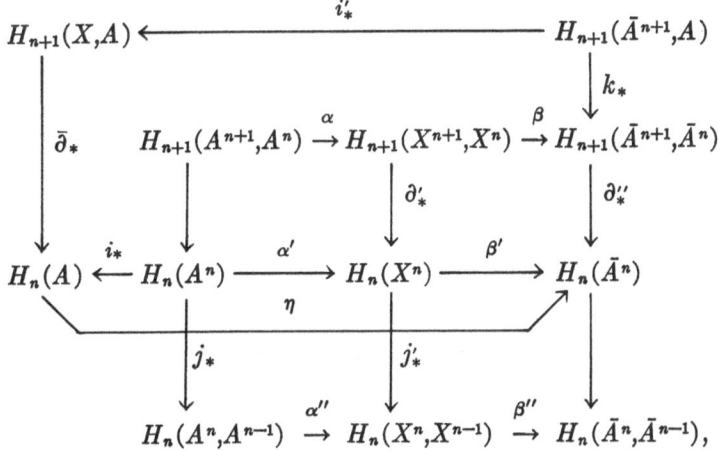

Now $H_{n+1}(\bar{A}^n,A) \approx \tilde{H}_{n+1}((X^n \cup A)/A) \approx \tilde{H}_{n+1}(X^n/A^n) = 0$ by corollary 1.9, since X^n/A^n is an n-dimensional complex. From the exact sequence of the pair $(\bar{A}^n,A)$ we conclude that η is a monomorphism, and hence that $\Psi_n \partial_*(z) = \bar{\partial}_* \Psi_{n+1}(z)$. ∎

We now collect these propositions into the following.

THEOREM 2.13. *On the category of CW complexes there is a natural equivalence from the cellular homology theory to the singular homology theory.* ∎

Because of this equivalence, no confusion will result if we adopt the following notation. If $(X,\mathcal{S})$ is a CW complex, we denote its cellular homology by $H_*(X)$, and retain $C_*(\mathcal{S})$ for its cellular chain complex. We will use $C_*(X)$ for the singular chain complex of X.

Note that $\psi_0: Z_0(\mathcal{S}) \to Z_0(X)$ is a map of $C_0(\mathcal{S})$ to $C_0(X)$. It is the linear map which assigns to each 0-cell σ^0 of $C_0(\mathcal{S})$ the corresponding unique 0-simplex of $C_0(X)$.

Let $\epsilon': C_0(\mathcal{S}) \to R$ be the map $\epsilon'(\sum_\lambda a_\lambda \sigma_\lambda{}^0) = \sum_\lambda a_\lambda$, the *natural augmentation* on $C_0(\mathcal{S})$. Let ϵ be the standard augmentation of $C_0(X)$.

PROPOSITION 2.14. *If* $(X,\mathcal{S})$ *is a CW complex*
(i) *for any 0-chain* $c^0 \in C_0(\mathcal{S})$, *we have* $\epsilon\psi_0(c^0) = \epsilon'(c^0)$;
(ii) *for any 1-chain* $c^1 \in C_1(\mathcal{S})$, *we have* $\epsilon'\partial(c^1) = 0$.

Proof. Statement (i) is obvious. To prove (ii), we need only prove it for generators σ of $C_1(\mathcal{S})$. Let z be a representative cycle for the class σ of $C_1(\mathcal{S}) = H_1(X^1,X^0)$, and let z' be any 1-chain of $C_1(X^1)$ which maps onto z. Then $\partial(z') \in Z_0(X^0) = Z_0(X^0,X^{-1})$ is a representative of $\partial(\sigma) \in C_0(\mathcal{S})$. But $\epsilon\partial(z') = 0$, because the augmentation of a bounding singular cycle is zero. Thus $\epsilon'\partial(\sigma) = \epsilon\psi_0\partial(\sigma) = \epsilon\partial(z') = 0$. ∎

By the usual argument we obtain the following.

COROLLARY **2.15.** *The augmentation ϵ' induces a map $\epsilon'_*:H_0(\mathcal{S})\to R$.* ∎

THEOREM **2.16.** *The natural equivalence Ψ from the cellular to singular homology of a CW complex restricts to a natural equivalence of reduced homology theories. There is a natural isomorphism of short exact sequences*

$$\begin{array}{ccccccccc} 0 & \to & \tilde{H}_0(\mathcal{S}) & \to & H_0(\mathcal{S}) & \xrightarrow{\epsilon'_*} & R & \to & 0 \\ & & \downarrow \Psi_0 & & \downarrow \Psi_0 & & \downarrow = & & \\ 0 & \to & \tilde{H}_0(X) & \to & H_0(X) & \xrightarrow{\epsilon_*} & R & \to & 0. \end{array}$$

Proof. Commutativity in the right-hand square follows from the fact that on the chain level we have $\epsilon' = \epsilon\psi_0$. Thus $\Psi_0 \mid \tilde{H}_0(\mathcal{S})$ is an isomorphism of Ker ϵ'_* onto Ker ϵ_*. ∎

We can use the generators $\{j_\# 1_n\}$ of $H_n(\Delta^n,\dot{\Delta}^n)$ selected in section 1 to describe the equivalence Ψ of 2.13 more concretely. Fix a set of homeomorphisms $h_n:(\Delta^n,\dot{\Delta}^n)\to(D^n,\dot{D}^n)$ for $n=0$, 1, 2, $\cdots$. Let $(X,\mathcal{S})$ be a CW complex, and σ an n-cell of X. To each n-cell σ we will associate the singular n-simplex $\varphi_\sigma h_n:\Delta^n\to X$, which we denote by σ'. Let $j_\#:C_*(X^n)\to C_*(X^n,X^{n-1})$ be the quotient map. The singular chain σ' is not a cycle, but its boundary $\partial(\varphi_\sigma h_n) = \sum_i(-1)^i\varphi_\sigma h_n d_i^*$ is a sum of singular simplexes of X^{n-1}, so that $j_\#(\sigma')$ is a cycle of $C_n(X^n,X^{n-1})$. The class $\{j\varphi_\sigma h_n\}=\{\sigma''\}$ will be a generator of the direct summand $i_{\sigma*}(H_n(\sigma,\dot{\sigma}))\subset H_n(X^n,X^{n-1})$. These generators form a basis for the free R-module $H_n(X^n,X^{n-1})$, but of course not a unique one. The generators $\{\sigma''\}$ depend on the choice of the homeomorphisms h_n as well as on the characteristic maps φ.

We call a singular chain $c_p\in C_p(X^q)$ *secondary* if $p>q$. A secondary cycle, for example, is always a bounding cycle by 1.9.

PROPOSITION **2.17.** *Suppose $z=\sum_\lambda a_\lambda\sigma_\lambda$ is a cellular n-cycle of $C_n(\mathcal{S})$. Then $\Psi_n(\{z\})$ has a representative singular cycle $z''=$*

$\sum_\lambda a_\lambda \sigma'_\lambda + w$, where w is a secondary chain chosen to make $\partial(z'') = 0$. Thus, given a cellular cycle, we can easily write the "principal part" of a corresponding singular cycle.

Proof. Let $z' = \sum_\lambda a_\lambda \sigma'_\lambda$ be a chain of $C_n(X^n)$ which j maps onto z. Then $\partial(z') \in C_{n-1}(X^{n-1})$ is a representative of $\partial_*(\{z\})$, where $\partial_*: H_n(X^n, X^{n-1}) \to H_n(X^{n-1})$ is the boundary operator in the homology sequence of the pair (X^n, X^{n-1}). Since $i_*: H_{n-1}(X^{n-1}) \to H_n(X^{n-1}, X^{n-2})$ is a monomorphism, and since we are assuming $\partial(z) = i_* \partial_*(z) = 0$, it follows that $\partial(z')$ is a bounding cycle of $C_{n-1}(X^{n-1})$, say $\partial(z') = -\partial(w)$ for some $w \in C_n(X^{n-1})$. Thus $z'' = \sum_\lambda a_\lambda \sigma_\lambda + w$ is a cycle in $C_n(X^n)$ and in $C_n(X)$. By definition of Ψ, a representative of $\Psi(\{z\})$ is the image under $i_\#: C_n(X^n) \to C_n(X)$ of any cycle which $j_\#: C_n(X^n) \to C_n(X^n, X^{n-1})$ maps onto z. Clearly z'' is such a cycle, and the homology class of z'' in $C_n(X)$ is $\Psi(\{z\})$. ∎

We wish to make some remarks about continuous maps between CW complexes. If (X,A) and (Y,B) are CW pairs and $f: (X,A) \to (Y,B)$ is a continuous map, two applications of theorem II.8.5 yield a cellular map $f_0: (X,A) \to (Y,B)$ which is homotopic to f. If $f_1: (X,A) \to (Y,B)$ is any other cellular map homotopic to f, then f_0 and f_1 are homotopic. By corollary II.8.8, f_0 and f_1 are homotopic via a cellular homotopy, and $f_{0*} = f_{1*}: H_n(\bar{A}^n, \bar{A}^{n-1}) \to H_n(\bar{B}^n, \bar{B}^{n-1})$. Thus each continuous map $(X,A) \to (Y,B)$ induces a unique cellular chain map. This is a partial reason for defining the CW category to be the category whose objects are CW complexes and whose morphisms are all continuous maps between CW complexes.

It should be pointed out that the results of these first two sections can lead to a large simplification in the problem of calculation of the singular homology of a CW complex $(X, \mathcal{S})$ at least if X is compact; for if X is compact, $\mathcal{S}$ is a finite set of cells and the cellular chain modules are finitely generated free modules. Thus our problem is to compute the boundary operator in $C(\mathcal{S})$.

At this point we are able to do some homological calculations for some special CW complexes, in which this boundary operator is trivial.

LEMMA **2.18.** *Let X be a CW complex which has cells in dimension q indexed by Λ. Suppose that X has no cells in dimension $q+1$ or $q-1$. Then $H_q(X)=F_\Lambda$, where F_Λ is a free R-module with basis Λ, and $H_{q-1}(X)=H_{q+1}(X)=0$.*

Proof. Since $C_{q-1}(\mathcal{S})=C_{q+1}(\mathcal{S})=0$, $Z_{q-1}(\mathcal{S})=0=Z_{q+1}(\mathcal{S})$, which implies $H_{q-1}(X)=H_{q-1}(\mathcal{S})=0$ and $H_{q+1}(X)=H_{q+1}(\mathcal{S})=0$. Also $C_q(\mathcal{S})=Z_q(\mathcal{S})$ and $B_q(\mathcal{S})=0$, so that $H_q(X)\approx H_q(\mathcal{S})=Z_q(\mathcal{S})=C_q(\mathcal{S})=F_\Lambda$. ∎

Example **2.19.** Complex projective space. Using the decomposition of complex projection space given in I.2.4, $H_{2q}(\mathbf{CP}^n)\approx R$ for $q=0, 1, \cdots, n$, and $H_p(\mathbf{CP}^n)=0$ otherwise. Regarding $\mathbf{CP}^k\subset\mathbf{CP}^n$ for $k\leq n$ as in I.2.4, $H_{2q}(\mathbf{CP}^n/\mathbf{CP}^k)\approx R$ for $q=0$ or $q=k+1, k+2, \cdots, n$, and $H_p(\mathbf{CP}^n/\mathbf{CP}^k)=0$ otherwise.

Example **2.20.** Quaternionic projective space. Using the decomposition given in I.2.4, $H_{4q}(\mathbf{HP}^n)\approx R$ for $q=0, 1, \cdots, n$ and $H_p(\mathbf{HP}^n)=0$ otherwise.

Example **2.21.** Complex Grassman varieties $\mathbf{CG}(n,m)$. Using the description in example I.2.5, there is a decomposition of $\mathbf{CG}(n,m)$ which consists of cells of even dimension only. In fact, the number of cells of dimension $2q$ is $\rho_q(n,m)=$ card $\{(p_n, \cdots, p_1)\in Z^n \mid p_k\geq 0, 1\leq k\leq n, \sum_k p_k\leq m$, and $\sum_k kp_k=q\}$. From this and lemma 2.18 we obtain the following.

For the space $\mathbf{CG}(n,m)$:
$H_{2q}(\mathbf{CG}(n,m))$ is a free R module of rank $\rho_q(n,m)$
$H_{2q-1}(\mathbf{CG}(n,m))=0$.

Example **2.22.** Quaternionic Grassman varieties $\mathbf{HG}(n,m)$. Using the description of example I.2.5, a calculation analogous to the one above yields:

$H_{4q}(\mathbf{HG}(n,m))$ is a free R-module of rank $\rho_q(n,m)$,
$H_p(\mathbf{HG}(n,m)) = 0$ if $p \not\equiv 0 \pmod 4$.

3. ORIENTATION, INCIDENCE, AND DEGREE

To calculate cellular homology, we must choose generators for our free chain modules and express boundary operators and induced maps in terms of them.

Definition **3.1.** If $\varphi_\lambda : D^n \to \sigma_\lambda{}^n \subset X$ is a characteristic map for the n-cell $\sigma_\lambda{}^n$ of X, a choice of a generator of the direct summand $\varphi_{\lambda*}(H_n(D^n,\dot{D}^n)) \subset H_n(X^n,X^{n-1})$ is an *orientation of the cell* $\sigma_\lambda{}^n$. A choice of such generators for all cells of X is an *orientation of* X. A CW complex X is *oriented* if an orientation has been chosen for X. It should be noted that any orientation of X may be written as

$$\{\varphi_{\lambda*}(u_\lambda \zeta_n) \mid \lambda \in \Lambda_n,\ u_\lambda \text{ is a unit of } R, \text{ and } n = 0,1, \cdots\}.$$

It is conventional to denote the orientation of the cell $\sigma_\lambda{}^n$ by $\sigma_\lambda{}^n \in H_n(X^n,X^{n-1})$, i.e., $\sigma_\lambda{}^n$ is used both for the geometric cell $\sigma_\lambda{}^n \subset X$ and an algebraic object $\sigma_\lambda{}^n \in H_n(X^n,X^{n-1})$, in which case it is called an (algebraic) n-cell. It should be clear from the context which usage is intended, but, if not, we will specify the meaning of $\sigma_\lambda{}^n$.

If $(X,\mathcal{S})$ is an oriented CW complex and $c \in C_n(\mathcal{S}) = H_n(X^n,X^{n-1})$, we write $c = \sum c_\lambda \sigma_\lambda{}^n$, where $c_\lambda \in R$ for all $\lambda \in \Lambda_n$ and $c_\lambda = 0$ for all but finitely many $\lambda \in \Lambda_n$.

Definition **3.2.** If $(X,\mathcal{S})$ is an oriented CW complex and $\partial_n : C_n(\mathcal{S}) \to C_{n-1}(\mathcal{S})$ is the boundary homomorphism, then for each n-cell $\sigma^n \in C_n(\mathcal{S})$,

$$\partial \sigma^n = \sum_{\lambda \in \Lambda_{n-1}} [\sigma^n : \sigma_\lambda{}^{n-1}] \sigma_\lambda{}^{n-1},$$

where $[\sigma^n : \sigma_\lambda{}^{n-1}] \in R$. The ring element $[\sigma^n : \sigma_\lambda{}^{n-1}]$ is the *incidence number* of σ^n and $\sigma_\lambda{}^{n-1}$. We say that the (algebraic) cells σ^n and σ^{n-1} are incident if $[\sigma^n:\sigma^{n-1}]\neq 0$.

Proposition 3.3. *Let X be an oriented CW complex, let σ^n and τ^{n-1} be cells of X, and let $j_\tau: X^{n-1}/X^{n-2}\to\tau/\dot\tau$ be the map which collapses all other spheres of X^{n-1}/X^{n-2} to the base point. Let $f(\sigma:\tau)$ be the composition*

$$\dot D^n \xrightarrow{\varphi_\sigma} X^{n-1} \xrightarrow{p} X^{n-1}/X^{n-2} \xrightarrow{j_\tau} \tau/\dot\tau.$$

If $u_\sigma\xi$ is the generator of $H_{n-1}(\dot D^n)$ determined by the orientation of σ, and τ' the generator of $H_{n-1}(\tau/\dot\tau)$ determined by the orientation of τ, then $f(\sigma:\tau)_*(u_\sigma\xi)=[\sigma:\tau]\tau'$.

Proof. Consider the commutative diagram

$$\begin{array}{ccccccc}
H_n(D^n,\dot D^n) & \xrightarrow{\partial_*} & H_{n-1}(\dot D^n) & & & & \\
\downarrow{\scriptstyle \varphi_{\sigma*}} & & \downarrow{\scriptstyle \varphi'_{\sigma*}} & & & & \\
H_n(X^n,X^{n-1}) & \xrightarrow{\partial_*} & H_{n-1}(X^{n-1}) & \xrightarrow{i_*} & H_{n-1}(X^{n-1},X^{n-2}) & \xrightarrow{j_*} & H_{n-1}(X^{n-1},X^{n-1}-(\tau-\dot\tau)) \\
 & & & \searrow{\scriptstyle p_*} & \downarrow{\scriptstyle \theta} & & \downarrow{\scriptstyle \theta} \\
 & & & & H_{n-1}(X^{n-1}/X^{n-2}) & \xrightarrow{j_{\tau*}} & H_{n-1}(\tau/\dot\tau),
\end{array}$$

where θ is the map of 1.3. Clearly, $\theta j_*\partial(\sigma)=[\sigma:\tau]\tau'$, so that if $\sigma=\varphi_{\sigma*}(u_\sigma\xi)$, then $[\sigma:\tau]\tau'=j_{\tau*}\theta i_*\varphi'_{\sigma*}(u_\sigma\xi)=f(\sigma:\tau)_*(u_\sigma\xi)$. ∎

The following corollary shows that if cells σ and τ of the oriented CW complex X are algebraically incident, $[\sigma:\tau]\neq 0$, then they are geometrically incident, $\tau\subset\dot\sigma$.

Corollary 3.4. *If σ^n and τ^{n-1} are cells of the oriented CW complex X and $\varphi_\sigma(\dot D^n)$ does not contain τ, then $[\sigma:\tau]=0$. Equivalently, if τ is not contained in $\dot\sigma$, then $[\sigma:\tau]=0$.*

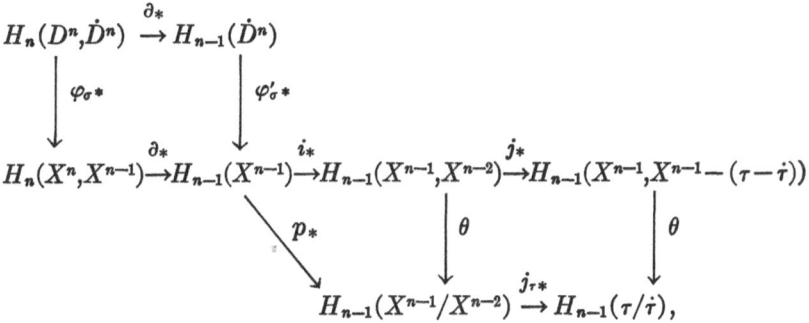

Proof. If $\varphi_\sigma(\dot{D}^n)$ does not contain τ, then $f(\sigma:\tau)$ is a map of $\mathbf{S}^{n-1}$ to $\mathbf{S}^{n-1}$ whose image omits a point. By radial deformation from this point, $f(\sigma:\tau)$ may be deformed to a constant map. Thus $f(\sigma:\tau)_*(u_\sigma\xi)=[\sigma:\tau]\tau=0$, so that $[\sigma:\tau]=0$. ∎

LEMMA 3.5. *Let $\bar{E}$ be a homeomorphic image of a closed n-cell in $\mathbf{S}^n$ whose interior E is an open n-cell. Let $f:\mathbf{S}^n\to\mathbf{S}^n$ map $\mathbf{S}^n-E$ onto a point $x\in\mathbf{S}^n$ and map E homeomorphically onto $\mathbf{S}^n-\{x\}$. Then $f_*:H_n(\mathbf{S}^n)\to H_n(\mathbf{S}^n)$ is an isomorphism.*

Proof. Let $h:D^n\to\bar{E}$ be a homeomorphism, let $y\in D^n$ be an interior point, and let $U=\mathbf{S}^n-\{h(y)\}$. Let F be a strong deformation retraction of $D^n-\{y\}$ onto $\dot{D}^n$, and define $\bar{F}:U\times I\to U$ by $\bar{F}(x,t)=x$ for $x\in\mathbf{S}^n-E$ and $\bar{F}(h(z),t)=hF(z,t)$ for $z\in D^n$. Then $\bar{F}$ is a strong deformation retraction of $U=\mathbf{S}^n-\{h(y)\}$ onto $\mathbf{S}^n-E$, so that the latter has trivial reduced homology, and the inclusion induces an isomorphism

$$i_*:H_n(\mathbf{S}^n)\xrightarrow{\approx}H_n(\mathbf{S}^n,\mathbf{S}^n-E).$$

Since $\mathbf{S}^n-E$ is a neighborhood deformation retract, an argument similar to that of 1.1 proves that the inclusion $e:(\bar{E},\dot{E})\to(\mathbf{S}^n,\mathbf{S}^n-E)$ induces homology isomorphisms. If $\bar{f}:(\bar{E},\dot{E})\to(\mathbf{S}^n,*)$ is the collapsing map, from the commutative diagram

$$\begin{array}{ccccc}
H_n(\mathbf{S}^n) & \underset{\approx}{\overset{i_*}{\longrightarrow}} & H_n(\mathbf{S}^n,\mathbf{S}^n-E) & \underset{\approx}{\overset{e_*}{\longleftarrow}} & H_n(E,\dot{E}) \\
\downarrow f_* & & \downarrow \bar{f}_* & \swarrow \bar{f}'_* & \\
H_n(\mathbf{S}^n) & \overset{\approx}{\longrightarrow} & H_n(\mathbf{S}^n,*) & &
\end{array}$$

and the fact that $\bar{f}'_*$ is an isomorphism, we conclude that f_* is an isomorphism. ∎

This lemma will enable us to prove a partial converse of 3.4.

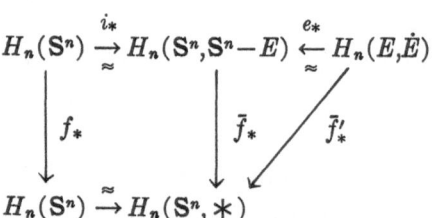

COROLLARY **3.6.** *If σ^n and τ^{n-1} are cells of the oriented CW complex X, if $\bar{E}$ is a closed $(n-1)$-cell in $\dot{D}^n$ whose interior is an open $(n-1)$-cell E, if $(\varphi_\sigma \mid \dot{D}^n)^{-1}(\tau - \dot{\tau}) = E$, and if φ_σ maps E homeomorphically onto $\tau - \dot{\tau}$, then $[\sigma:\tau]$ is a unit in R. In particular, this is the case for each pair of cells σ^n, τ^{n-1} of a regular cell complex for which $\tau^{n-1} \subset \sigma^n$.*

Proof. Under the conditions stated, the map $f(\sigma:\tau):\dot{D}^n \to \tau/\dot{\tau}$ induces an isomorphism

$$f(\sigma:\tau)_*:H_{n-1}(\dot{D}^n) \xrightarrow{\approx} H_{n-1}(\tau/\dot{\tau}).$$

Since $u_\sigma\xi$ is a generator of $H_{n-1}(\dot{D}^n)$, we see that $[\sigma:\tau]\tau' = f(\sigma:\tau)_*(u_\sigma\xi)$ is a generator of $H_{n-1}(\tau/\dot{\tau})$, so that $[\sigma:\tau]$ must be a unit of R. In the case of a regular complex, we need only choose homeomorphisms as characteristic maps of the cells in order to satisfy the hypotheses of the corollary. ∎

Definition **3.7.** Let $(X,\mathcal{S})$ and $(Y,\mathfrak{J})$ be oriented CW complexes and let $f:X \to Y$ be a cellular map. The cells of $\mathcal{S}$ will be denoted by $\sigma_\lambda{}^n$ and those of $\mathfrak{J}$ by $\tau_\mu{}^n$. If $f_\#:C_n(\mathcal{S}) \to C_n(\mathfrak{J})$ is the map induced by f, we write

$$f_\#(\sigma^n) = \sum_{\mu\in M_n}[f:\sigma^n:\tau_\mu{}^n]\tau_\mu{}^n,$$

where $[f:\sigma^n:\tau_\mu{}^n] \in R$ and $[f:\sigma^n:\tau_\mu{}^n] = 0$ for all but finitely many $\mu \in M_n$. The ring element $[f:\sigma^n:\tau_\mu{}^n]$ is the *degree* with which σ^n is mapped on $\tau_\mu{}^n$ by f.

We have a proposition similar to 3.3 about degrees of cellular maps. The proof, and that of the corresponding corollaries, is similar and will therefore be omitted.

PROPOSITION **3.8.** *Let X and Y be oriented CW complexes, let $f:X \to Y$ be a cellular map, let σ and τ be n-cells of X and Y, respectively, and let $F(\sigma:\tau)$ be the composite map*

$$F(\sigma:\tau):D^n/\dot{D}^n \xrightarrow{\bar{\varphi}_\sigma} X^n/X^{n-1} \xrightarrow{\bar{f}} Y^n/Y^{n-1} \xrightarrow{j_\tau} \tau/\dot{\tau},$$

where $\bar{\varphi}_\sigma$ *and* $\bar{f}$ *are induced by* φ_σ *and* f, *respectively, and* j_τ *is as in* 3.3. *If* $u_\sigma\xi_n$ *is the generator of* $H_n(D^n/\dot{D}^n)$ *corresponding to the orientation of* σ, *and* τ *is the generator of* $H_n(\tau/\dot{\tau})$ *corresponding to the orientation of* τ, *then* $F(\sigma:\tau)_*(u_\sigma\xi_n) = [f:\sigma:\tau]\tau$. ▮

COROLLARY **3.9.** *If* f,σ *and* τ *are as above, and if* τ *is not a subset of* $f(\sigma)$, *then* $[f:\sigma:\tau]=0$. ▮

COROLLARY **3.10.** *If* f,σ, *and* τ *are as above, and if* $f^{-1}(\tau-\dot{\tau})$ *is an open cell in* $\sigma-\dot{\sigma}$ *on which* f *is a homeomorphism, then* $[f:\sigma:\tau]$ *is a unit of* R. ▮

The following two propositions interpret the cellular chain complex in terms of incidence and cellular chain maps in terms of degree.

PROPOSITION **3.11.** *Let* $(X,\mathcal{S})$ *be an oriented CW complex.*

(i) *If* σ^{n-1} *is not contained in* σ^n, *then* $[\sigma^n:\sigma^{n-1}]=0$.

(ii) *If* σ^1 *is a 1-cell such that* $\dot{\sigma}^1$ *is a single vertex, then* $[\sigma^1:\sigma^0]=0$ *for all 0-cells* σ^0.

(iii) *If* σ^1 *is a 1-cell such that* σ^1 *consists of two vertices* $\sigma_\mu{}^0$, $\sigma_\nu{}^0$, μ, $\nu\in\Lambda_0$, *then* $[\sigma^1:\sigma_\mu{}^0]=-[\sigma^1:\sigma_\nu{}^0]$ *is a unit of the ring* R *and* $[\sigma^1:\sigma_\lambda{}^0]=0$ *for* $\lambda\neq\mu,\nu$.

(iv) *If* $n\geq 2$, *and* σ^n *and* σ^{n-2} *are cells of* $\mathcal{S}$, *then*

$$\textstyle\sum_{\lambda\in\Lambda_{n-1}}[\sigma^n:\sigma_\lambda{}^{n-1}][\sigma_\lambda{}^{n-1}:\sigma^{n-2}]=0.$$

Proof. Statement (i) is just 3.4. Statements (ii) and (iii) follow because in the augmented chain module $C_*(\mathcal{S})$, $\epsilon\partial=0$. Since the cells $\{\sigma_\lambda\}$ are a basis for $C_*(\mathcal{S})$ and $\partial\partial=0$, we see that (iv) holds. ▮

PROPOSITION **3.12.** *If* $(X,\mathcal{S})$ *and* $(Y,\mathfrak{T})$ *are oriented CW complexes and* $f:X\rightarrow Y$ *is a cellular map, the degrees of* f *satisfy the following conditions.*

(i) *If* $\tau_\mu{}^n$ *is not a subset of* $f(\sigma^n)$, *then* $[f:\sigma^n:\tau_\mu{}^n]=0$.

(ii) *If* $f(\sigma^0)=\tau^0$, *then* $[f:\sigma^0:\tau^0]=1$.

(iii) *For* $n \geq 1$ *and* $\sigma^n \in \mathcal{S}$ *and* $\tau^{n-1} \in \mathcal{T}$,

$$\sum_{\mu \in M_n} [f:\sigma^n:\tau_\mu{}^n][\tau_\mu{}^n:\tau^{n-1}] = \sum_{\lambda \in \Lambda_{n-1}} [\sigma^n:\sigma_\lambda{}^{n-1}][f:\sigma_\lambda{}^{n-1}:\tau^{n-1}].$$

Proof. Statement (i) is corollary 3.9. Statements (ii) and (iii) express the fact that $f_\#$ is a chain map which preserves augmentation. ∎

We remark that these conditions are *not* sufficient to determine the incidence numbers of a CW complex or the degrees of a cellular map.

For example, let R be the ring of integers and consider the following CW complexes $(X_i,\mathcal{S}_i)$, $i=1,2$, where $\mathcal{S}_i = \{\sigma^0,\sigma^1,\sigma^2\}$. The space X_1 is a 2-disc D^2 with the decomposition of I.2.1 (see Fig. 5.1). Incidence numbers for $(X_1,\mathcal{S}_1)$ are $[\sigma^1:\sigma^0]=0$ and

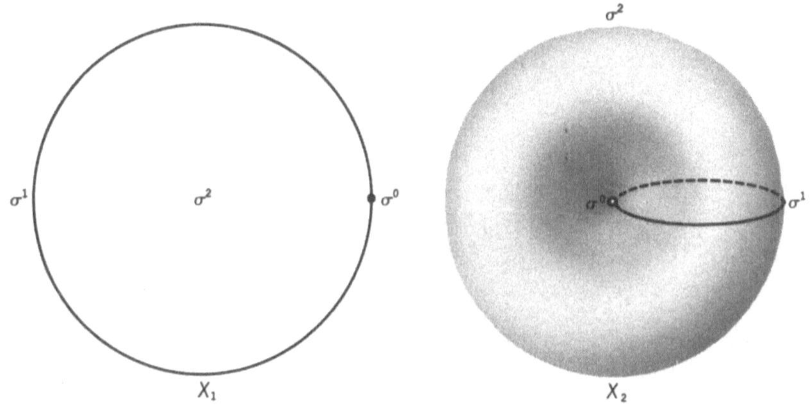

Figure 5.1

$[\sigma^2:\sigma^1]=1$; consequently the homology groups are $H_0(X_1) \approx R$ and $H_k(X_1)=0$ for $k>0$. The space X_2 is a 2-sphere with exactly two antipodal points identified. The cellular structure may be described by giving $\mathbf{S}^2$ a cellular structure with two 0-cells $\sigma_1{}^0,\sigma_2{}^0$; a 1-cell σ^1 with $\dot{\sigma}^1 = \sigma_1{}^0 \cup \sigma_2{}^0$; a 2-cell σ^2 attached by projecting $\dot{D}^2$ onto D^1 by $(x,y) \to x$ and then using the characteristic map of σ^1; and finally identifying $\sigma_1{}^0$ and $\sigma_2{}^0$. Incidence numbers for $(X_2,\mathcal{S}_2)$ are $[\sigma^1:\sigma^0]=[\sigma^2:\sigma^1]=0$, so that $H_0(X_2) \approx H_1(X_1) \approx H_2(X_2) \approx R$, and $H_k(X_2)=0$ for $k>2$.

There are cellular maps from $(\mathbf{S}^1, *)$ to $(\mathbf{S}^1, *)$ having any degree on a 1-cell. For example, if we identify $\mathbf{S}^1$ with the complex numbers of norm 1 and let $* = \sigma^0 = \{1\}$ and $\sigma^1 = \mathbf{S}^1$, then the map $f_m: (\mathbf{S}^1, *) \to (\mathbf{S}^1, *)$ defined by $f_m(z) = z^m$ has degree m on the 1-cell.

4. REGULAR CW COMPLEXES AND PROPER MAPS

We have seen that the incidence relations are not sufficient to determine the incidence numbers of an arbitrary CW complex, and that the degree relations are not sufficient to determine the degrees of a cellular map. The case of regular complexes, however, is different. Our major results are 4.3 and 4.8, which show that there is a finite method for calculating the homology of a regular complex and, under suitable conditions, the homomorphism induced by a cellular map.

Recall from III.1.5 that if σ and τ are cells of a regular CW complex, τ is a face of σ if $\tau \subset \sigma$.

LEMMA **4.1.** *If $(X, \mathcal{S})$ is a regular CW complex and $\sigma^{n-2} \subset \sigma^n$ are cells of X, then there are exactly two $(n-1)$-cells $\sigma_1^{n-1}, \sigma_2^{n-1}$ satisfying $\sigma^{n-2} \subset \sigma_i^{n-1} \subset \sigma^n$, for $i = 1, 2$.*

Proof. Since X is regular, $\dot{\sigma}^n$ is a subcomplex of X which is a topological $(n-1)$-manifold, in fact an $(n-1)$-sphere. By proposition III.1.4, $\dot{\sigma}^n$ is a regular CW complex. By theorem III.2.2 (ii), each $(n-2)$-cell is a face of exactly two $(n-1)$-cells. ∎

The main result about regular CW complexes is the following one.

THEOREM **4.2.** *Suppose $(X, \mathcal{S})$ is a regular CW complex. Let Λ_k be an indexing set for the k-cells of X, and consider the following conditions on the set*

$$\mathcal{A} = \{a^n_{\lambda,\mu} \in R \mid \lambda \in \Lambda_n, \mu \in \Lambda_{n-1}, n = 0, 1, \cdots\}.$$

(i) *If* σ_μ^{n-1} *is not a face of* σ_λ^n, *then* $a^n_{\lambda,\mu}=0$.
(ii) *If* σ_μ^{n-1} *is a face of* σ_λ^n, *then* $a^n_{\lambda,\mu}$ *is a unit of* R.
(iii) *If* σ_λ^1 *is a* 1-*cell with vertices* $\sigma^0_{\mu 1}$ *and* $\sigma^0_{\mu 2}$, *then* $a^1_{\lambda,\mu 1}+a^1_{\lambda,\mu 2}=0$.
(iv) *If* $n\geq 2$ *and* $\sigma^{n-1}_{\mu 1}$, $\sigma^{n-1}_{\mu 2}$ *are the unique* $(n-1)$-*cells such that* $\sigma_\nu^{n-2}\subset\sigma^{n-1}_{\mu 1}\subset\sigma_\lambda^n$ *and* $\sigma_\nu^{n-2}\subset\sigma^{n-1}_{\mu 2}\subset\sigma_\lambda^n$, *then*

$$a^n_{\lambda,\mu 1}a^{n-1}_{\mu 1,\nu}+a^n_{\lambda,\mu 2}a^{n-1}_{\mu 2,\nu}=0.$$

Then the incidence numbers for any orientation of X *satisfy these conditions* $a^n_{\lambda,\mu}=[\sigma_\lambda^n:\sigma_\mu^{n-1}]$. *Conversely, given any such* $\mathfrak{a}$, *there is a unique choice of orientations for the cells of* $\mathcal{S}$ *such that* $[\sigma_\lambda^n:\sigma_\mu^{n-1}]=a^n_{\lambda,\mu}$.

The theorem states that if we write any differential operator ∂ on the module $C_*(\mathcal{S})$ satisfying (i), (ii), (iii), and (iv), then the corresponding homology will be $H_*(X)$. In the proof we will see that given any such system of incidence numbers constructed in dimensions $\leq n-1$, there is always an extension to dimension n. This completely resolves the problem of finding the differential operator in the cellular chains of a *regular* complex.

Proof. Let X be an oriented regular CW complex and suppose that $a^n_{\lambda,\mu}=[\sigma_\lambda^n:\sigma_\mu^{n-1}]$. Then (i) and (iii) follow from 3.11, and (iv) follows from 3.11 and 4.1. Statement (ii) follows from corollary 3.6.

The proof of the converse will be somewhat longer. We orient the cells by induction on their dimension.

The 0-cells are oriented uniquely by choosing $\sigma=\varphi_{\sigma *}(\zeta_0)$, where $\varphi_\sigma=D^0\to X$ is a characteristic map for the 0-cell. Note that from the diagram

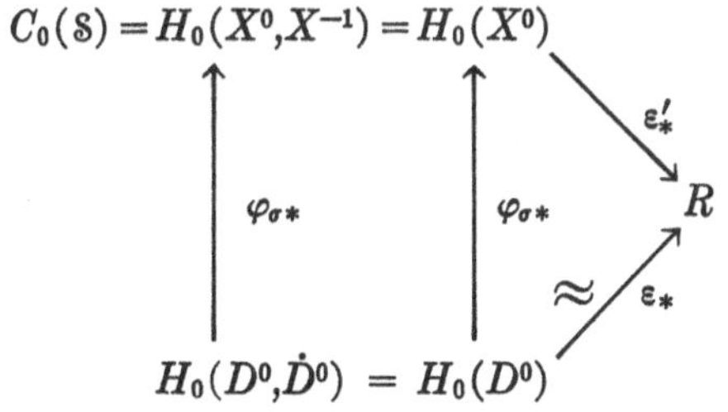

we can infer $\epsilon'_*(\sigma)=\epsilon'_*\varphi_{\sigma *}(\zeta_0)=\epsilon_*(\zeta_0)=1\in R$.

To orient the 1-cell σ_λ^1, let φ^1, φ_1^0, and φ_2^0 be characteristic homeomorphisms for $\sigma_\lambda{}^1$, $\sigma^0_{\mu 1}$ and $\sigma^0_{\mu 2}$, respectively, where $\sigma^0_{\mu 1}$, $\sigma^0_{\mu 2}$ are vertices of $\sigma_\lambda{}^1$. Let $\varphi_\pm : D^0 \to \dot{D}^1$ be the maps $\varphi_+(0) = +1$, $\varphi_-(0) = -1$. If $\varphi_1{}^0 = \varphi^1\varphi_+$, define $\sigma_\lambda{}^1 \in C_1(\mathcal{S})$ by $\sigma_\lambda{}^1 = \varphi^1_*(a^1_{\lambda,\mu 1}\zeta_1)$. From the commutative diagram

$$\begin{array}{ccccc} H_1(X^1,X^0) & \xrightarrow{\partial_*} & H_0(X^0) & & \\ \uparrow \varphi^1_* & & \uparrow \varphi^1_* & \nwarrow \varphi^0_{1*} & \\ H_1(D^1,\dot{D}^1) & \xrightarrow{\partial'_*} & H_0(\dot{D}^1) & \xleftarrow{\varphi_{+*}} & H_0(D^0) \end{array}$$

we see that

$$[\sigma_\lambda{}^1 : \sigma^0_{\mu 1}]\sigma^0_{\mu 1} + [\sigma_\lambda{}^1 : \sigma^0_{\mu 2}]\sigma^0_{\mu 2} = \partial_*\sigma_\lambda{}^1 = \partial_*\varphi^1_*(a^1_{\lambda,\mu 1}\zeta_1) = a^1_{\lambda,\mu 1}\varphi^1_*\partial_*\zeta_1$$

$$= a^1_{\lambda,\mu 1}(\varphi^1_*\varphi_{+*}(\zeta_0) - \varphi^1_*\varphi_{-*}(\zeta_0))$$

$$= a^1_{\lambda,\mu 1}\sigma^0_{\mu 1} - a^1_{\lambda,\mu 2}\sigma^0_{\mu 2}.$$

Thus

$$[\sigma_\lambda{}^1 : \sigma^0_{\mu 1}] = a^1_{\lambda,\mu 1} \text{ and } [\sigma_\lambda{}^1 : \sigma^0_{\mu 2}] = -a^1_{\lambda,\mu 1} = a^1_{\lambda,\mu 2}.$$

This is clearly the only choice of orientation for $\sigma_\lambda{}^1$ with the proper incidence numbers.

Now suppose that all cells of dimension k, $1 \leq k \leq n-1$ have been uniquely oriented so that $[\sigma_\lambda{}^k : \sigma_\mu{}^{k-1}] = a^k_{\lambda,\mu}$, and let $\sigma_\lambda{}^n$ be an n-cell. If $\alpha = \sum_\mu a^n_{\lambda,\mu}\sigma_\mu{}^{n-1}$, then

$$\partial\alpha = \sum_\mu a^n_{\lambda,\mu} \sum_\nu [\sigma_\mu{}^{n-1} : \sigma_\nu{}^{n-2}]\sigma_\nu{}^{n-2} = \sum_\nu \sum_\mu a^n_{\lambda,\mu} a^{n-1}_{\mu,\nu}\sigma_\nu{}^{n-2}$$

since we are assuming $[\sigma_\mu{}^{n-1} : \sigma_\nu{}^{n-2}] = a^{n-1}_{\mu,\nu}$. By property (d), $\partial\alpha = 0$, so that α is a cycle on the subcomplex $\dot{\sigma}_\lambda{}^n \subset X$. Let $\beta \in H_n(X^n, X^{n-1})$ be an arbitrary orientation for $\sigma_\lambda{}^n$, so that $\partial\beta = \sum_\mu [\beta : \sigma_\mu{}^{n-1}]\sigma_\mu{}^{n-1}$ is an $(n-1)$-cycle on $\dot{\sigma}_\lambda{}^n$. Now $Z_{n-1}(\dot{\sigma}_\lambda{}^n) = H_{n-1}(\dot{\sigma}_\lambda{}^n) \approx R$ because $B_{n-1}(\dot{\sigma}_\lambda{}^n) = 0$. Let $\gamma = \sum_\mu c_\mu\sigma_\mu{}^{n-1}$ ($c_\mu = 0$ unless $\sigma_\mu{}^{n-1} \subset \sigma_\lambda{}^n$) be a generating cycle for $Z_{n-1}(\dot{\sigma}_\lambda{}^n)$. Then $\alpha = a\gamma$ and $\partial\beta = b\gamma$ for $a, b \in R$. Since the $\sigma_\mu{}^{n-1}$ are part of a basis for $C_{n-1}(\mathcal{S})$, we see that $a^n_{\lambda,\mu} = ac_\mu$ and $[\beta : \sigma_\mu{}^{n-1}] = bc_\mu$ for all μ. But $[\beta : \sigma_\mu{}^{n-1}]$ is a unit in R, so that b is a unit in R, and $a^n_{\lambda,\mu} = ab^{-1}[\beta : \sigma_\mu{}^{n-1}]$ for all μ. If we let $\sigma_\lambda{}^n = ab^{-1}\beta \in H_n(X^n, X^{n-1})$, then

$$\partial\sigma_\lambda{}^n = \sum_\mu [\sigma_\lambda{}^n : \sigma_\mu{}^{n-1}]\sigma_\mu{}^{n-1} = \sum_\mu a^n_{\lambda,\mu}\sigma_\mu{}^{n-1}.$$

If $\alpha \in H_n(X^n, X^{n-1})$ is a new orientation for the geometric cell $\sigma_\lambda{}^n$, then $\alpha = u\sigma_\lambda{}^n$ for some unit $u \in R$, and $\partial\alpha = \sum_\mu u a^n_{\lambda,\mu}\sigma_\mu{}^{n-1}$. By the induction hypothesis, the orientations $\sigma_\mu{}^{n-1}$ are uniquely determined by the numbers $\mathfrak{a}$, so that the incidence numbers for α are $ua^n_{\lambda,\mu}$. Since X is regular, we may choose some $a^n_{\lambda,\mu} \neq 0$, and we see that α is an orientation with the proper incidence numbers only if $u = 1$, i.e., $\alpha = \sigma^n_{\lambda,\mu}$.

Doing this for each n-cell of X, we orient X^n uniquely, and by induction there is a unique orientation of X with incidence numbers $\mathfrak{a}$. ∎

As an application of 4.2, we state the following two theorems without proof.

THEOREM 4.3. *Let $(X, \mathcal{S})$ and $(Y, \mathfrak{J})$ be regular oriented CW complexes. Then $(X \otimes Y, \mathcal{S} \otimes \mathfrak{J})$ is a (regular) CW complex and there is an orientation of $(X \otimes Y, \mathcal{S} \otimes \mathfrak{J})$ such that*

(a) $[\sigma_1{}^n \times \tau_1{}^m : \sigma_2{}^p \times \tau_2{}^q] = 0$ *if* $\sigma_1{}^n \neq \sigma_2{}^p$ *and* $\tau_1{}^m \neq \tau_2{}^q$;

(b) $[\sigma^n \times \tau^m : \sigma^{n-1} \times \tau^m] = [\sigma^n : \sigma^{n-1}]$;

(c) $[\sigma^n \times \tau^m : \sigma^n \times \tau^{m-1}] = (-1)^n [\tau^m : \tau^{m-1}]$. ∎

We remark that by changing the descriptions (a), (b), and (c), we can obtain other orientations of $(X \otimes Y, \mathcal{S} \otimes \mathfrak{J})$, but the one we have given has become standard. In 6.1, we will prove a generalization of 4.3 for arbitrary CW complexes.

THEOREM 4.4. *Let $(X, \mathcal{S})$ be a CW simplicial complex with vertices in the partially ordered set V such that each simplex has linearly ordered vertices. Let $\langle v_{i_0}, v_{i_1}, \cdots, v_{i_n}\rangle$, $i_0 < i_1 < \cdots < i_n$, denote the n-simplex with vertices $v_{i_0}, v_{i_1}, \cdots, v_{i_n}$. Define*

(a) $[\langle v_{i_0}, \cdots, v_{i_1}\rangle : \langle v_{j_0}, \cdots, v_{j_{n-1}}\rangle] = 0$ *if* $\{v_{j_0}, \cdots, v_{j_{n-1}}\}$ *is not a subset of* $\{v_{i_0}, \cdots, v_{i_n}\}$;

(b) $[\langle v_{i_0}, \cdots, v_{i_n}\rangle : \langle v_{i_0}, \cdots, \hat{v}_{i_k}, \cdots, v_{i_n}\rangle] = (-1)^k$ *for* $0 \leq k \leq n$, *and $\hat{v}$ means the vertex v is omitted.*

There is a unique orientation of the regular CW complex $(X, \mathcal{S})$ with incidence numbers $[\langle v_{i_0}, \cdots, v_{i_n}\rangle : \langle v_{j_0}, \cdots, v_{j_{n-1}}\rangle]$. ∎

If we note that the boundary operator described in this theorem is the usual simplicial boundary operator $\partial\langle v_{i_0}, \cdots, v_{i_n}\rangle =$

$\sum_{k=0}^{n}(-1)^k\langle v_{i_0}, \cdots, \hat{v}_{i_k}, \cdots, v_{i_n}\rangle$, we obtain the following classical result.

PROPOSITION 4.5. *If K is an abstract simplicial complex, then there is an isomorphism $\lambda: H'_*(K) \to H_*(|K|)$, where H' is the (oriented) simplicial homology of K.*

Proof. Take any vertex ordering on K so that the vertices of each simplex are linearly ordered. This orients $|K|$. Let $\mathcal{S}$ be the cell structure on the CW complex $|K|$. If $\sigma = \{v_0, \cdots, v_n\}$ is an n-simplex of K, let $\lambda'(\sigma)$ be the generator $\langle v_0, \cdots, v_n\rangle$ of $C_n(\mathcal{S})$. Then λ' is a chain isomorphism of the oriented simplicial chains $C'_*(K)$ with $C_*(\mathcal{S})$ which therefore induces an isomorphism of $H'(K)$ with $H_*(\mathcal{S})$. By 2.11 this is isomorphic to the singular homology $H_*(|K|)$. It is easy to check that the composition $\Psi \cdot \lambda' = \lambda$ is independent of the vertex ordering chosen. ∎

Definition 4.6. Let $(X, \mathcal{S})$ and $(Y, \mathfrak{J})$ be CW complexes, and let $f: X \to Y$ be a cellular map. The map f is *proper* if $\tilde{H}_n(C(f(\sigma^n))) = 0$ for all n and all $\sigma^n \in \mathcal{S}$, where $C(f(\sigma^n))$ is the carrier of $f(\sigma^n)$.

THEOREM 4.7. *Let $(X, \mathcal{S})$ and $(Y, \mathfrak{J})$ be oriented CW complexes, and let $f: X \to Y$ be a proper cellular map. Let Λ_k and M_k be indexing sets for the k-cells of X and Y, respectively, and consider the following conditions on the set $\mathcal{B} = \{b^n_{\lambda,\mu} \in R \mid \lambda \in \Lambda_n, \mu \in M_n, n = 0,1,2, \cdots\}$.*

(i) *If $\tau_\mu{}^n \in \mathfrak{J}$ is not a subset of $f(\sigma_\lambda{}^n)$, then $b^n_{\lambda,\mu} = 0$.*

(ii) *If $f(\sigma_\lambda{}^0) = \tau_\mu{}^0$, then $b^0_{\lambda,\mu} = 1$.*

(iii) *If $n \geq 1$, $\sigma_\kappa{}^n \in \mathcal{S}$, and $\tau_\nu{}^{n-1} \in \mathfrak{J}$, then*

$$\sum_\mu b^n_{\kappa,\mu}[\tau_\mu{}^n : \tau_\nu{}^{n-1}] = \sum_\lambda [\sigma_\kappa{}^n : \sigma_\lambda{}^{n-1}] b^{n-1}_{\lambda,\nu}.$$

Then the degrees of f satisfy these conditions with $b^n_{\lambda,\mu} = [f: \sigma_\lambda{}^n : \tau_\mu{}^n]$. Conversely, given any such collection $\mathcal{B}$, we have $b^n_{\lambda,\mu} = [f: \sigma_\lambda{}^n : \tau_\mu{}^n]$.

Proof. That the degrees of f satisfy the conditions (i), (ii), and (iii) is just proposition 3.12.

To prove the converse, let $a^n_{\lambda,\mu} = [f: \sigma_\lambda{}^n : \tau_\mu{}^n] \in R$, and let $b^n_{\lambda,\mu} \in \mathcal{B}$. We will prove that $a^n_{\lambda,\mu} = b^n_{\lambda,\mu}$ by induction on n.

For $n=0$, $b^0_{\lambda,\mu}=1$ if $f(\sigma_\lambda{}^0)=\tau_\mu{}^0$, and is zero otherwise by (i) and (ii). Thus $b^0_{\lambda,\mu}=a^0_{\lambda,\mu}$.

Suppose that $b^k_{\lambda,\mu}=a^k_{\lambda,\mu}$ for $0\leq k\leq n-1$, and let

$$\alpha_\kappa{}^n=\sum\nolimits_\mu (b^n_{\kappa,\mu}-a^n_{\kappa,\mu})\tau_\mu{}^n\in C_n(\mathfrak{Z}).$$

Using property (iii), we have

$$\begin{aligned}\partial\alpha_\kappa{}^n &= \sum\nolimits_{\mu,\nu} (b^n_{\kappa,\mu}-a^n_{\kappa,\mu})[\tau_\mu{}^n:\tau_\nu{}^{n-1}]\tau_\nu{}^{n-1}\\ &= \sum\nolimits_{\mu,\nu} (b^n_{\kappa,\mu}[\tau_\mu{}^n:\tau_\nu{}^{n-1}]-a^n_{\kappa,\mu}[\tau_\mu{}^n:\tau_\nu{}^{n-1}])\tau_\nu{}^{n-1}\\ &= \sum\nolimits_{\lambda,\nu} ([\sigma_\kappa{}^n:\sigma_\lambda{}^{n-1}]b^{n-1}_{\lambda,\nu}-[\sigma_\kappa{}^n:\sigma_\lambda{}^{n-1}]a^{n-1}_{\lambda,\nu})\tau_\nu{}^{n-1}\\ &= 0,\end{aligned}$$

since by the induction hypothesis $b^{n-1}_{\lambda,\nu}=a^{n-1}_{\lambda,\nu}$. Thus $\alpha_\kappa{}^n$ is a cycle in $C_n(\mathfrak{Z})$. By condition (i), $\alpha_\kappa{}^n$ is a cycle on the subcomplex $C(f(\sigma_\kappa{}^n))\subset Y^n$. Since $C(f(\sigma_\kappa))$ is a subcomplex of Y of dimension less than $n+1$, there are no nontrivial bounding n-cycles. Thus $\alpha_\kappa{}^n\in \tilde{H}_n(C(f(\sigma_\kappa{}^n)))=0$, so that $\alpha_\kappa{}^n=0$. Since the $\tau_\mu{}^n$ form a basis for $C_n(\mathfrak{Z})$, we must have $b^n_{\lambda,\mu}=a^n_{\lambda,\mu}$. By induction $a^n_{\lambda,\mu}=b^n_{\lambda,\mu}$ for all n. ∎

We remark that the composition of proper maps need not be proper. For example, let $X=D^1$ with the cell structure of I.2.2; let $Y=D^1$ with cells $\sigma_i{}^0=\{i-1\}$ for $i=0,1,2$, and 1-cells $\sigma_1{}^1=[-1,0]$ and $\sigma_2{}^1=[0,1]$; and let $Z=\mathbf{S}^1$ with the cell structure of I.2.2. Let $f:X\to Y$ be the identity, and

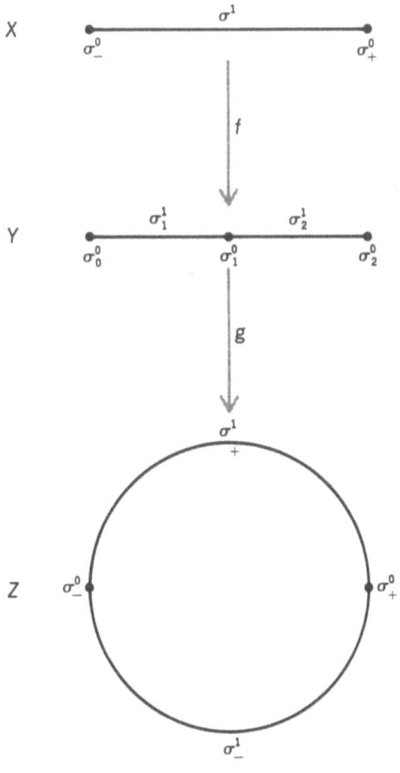

Figure 5.2

let $g:Y\to Z$ be the map $g(t)=e^{\pi i t}$. Then g and f are proper maps but gf is not (see Fig. 5.2).

The following is a criterion for proper maps.

PROPOSITION **4.8.** *Let $(X,\mathcal{S})$ be a CW complex, let $(Y,\mathfrak{J})$ be a regular CW complex, and let $f:X\to Y$ be cellular. If for each cell $\sigma^n\in\mathcal{S}$ there is an n-cell $\tau^n\in\mathfrak{J}$ such that $f(\sigma^n)\subset\tau^n\cup Y^{n-1}$, then f is a proper map.*

Proof. Since Y is regular, $\tau^n\cup Y^{n-1}$ is a subcomplex of Y, and $C(f(\sigma^n))\subset\tau^n\cup Y^{n-1}$. Let $\mathfrak{J}'$ be the set of cells of $C(f(\sigma^n))$. Since $\mathfrak{J}'$ contains no $(n+1)$-cells, $H_n(C(f(\sigma^n)))=Z_n(\mathfrak{J}')$. But $\mathfrak{J}'$ contains exactly one n-cell τ^n, so that every n-chain is a multiple of τ^n, and since Y is regular, τ^n has $(n-1)$-faces, so that $\partial\tau^n\neq 0$. Thus $H_n(C(f(\sigma^n)))=Z_n(\mathfrak{J}')=0$, and f is a proper map. ∎

COROLLARY **4.9.** *Let $(X,\mathcal{S})$ be a CW complex, let $(Y,\mathfrak{J})$ be a regular CW complex, and let $f:X\to Y$ be a regular cellular map. Then f is a proper cellular map.* ∎

Example **4.10.** Let $(X,\mathcal{S})$ be the regular CW complex on $X=\mathbf{S}^n$ given in I.2.2. Then $\mathcal{S}=\{\sigma_+{}^0,\sigma_-{}^0,\sigma_+{}^1,\cdots,\sigma_+{}^n,\sigma_-{}^n\}$, and $\dot\sigma_\pm{}^k=\sigma_+{}^{k-1}\cup\sigma_-{}^{k-1}$. For $k\geq 1$, define incidence numbers for $C_*(\mathcal{S})$ as follows:

$$[\sigma_+{}^k:\sigma_+{}^{k-1}]=1=(-1)^k[\sigma_+{}^k:\sigma_-{}^{k-1}]$$

$$[\sigma_-{}^k:\sigma_+{}^{k-1}]=1=(-1)^k[\sigma_-{}^k:\sigma_-{}^{k-1}].$$

Clearly conditions (i) through (iii) of theorem 4.2 are satisfied, and if $k\geq 2$, then

$$[\sigma_\pm{}^k:\sigma_\pm{}^{k-1}][\sigma_\pm{}^{k-1}:\sigma_\pm{}^{k-2}]+[\sigma_\pm{}^k:\sigma_\mp{}^{k-1}][\sigma_\mp{}^{k-1}:\sigma_\pm{}^{k-2}]$$
$$=1\cdot 1+(-1)^k(-1)^{k-1}=0,$$

$$[\sigma_\pm{}^k:\sigma_\pm{}^{k-1}][\sigma_\pm{}^{k-1}:\sigma_\mp{}^{k-2}]+[\sigma_\pm{}^k:\sigma_\mp{}^{k-1}][\sigma_\mp{}^{k-1}:\sigma_\mp{}^{k-2}]$$
$$=1\cdot(-1)^{k-1}+(-1)^k\cdot 1=0,$$

so that (iv) is satisfied as well. Let $(X,\mathcal{S})$ have the orientation determined by these incidence numbers.

Suppose that $\alpha \in C_k(\mathcal{S})$ for some k, $1 \leq k \leq n$. Then $\alpha = a\sigma_+{}^k + b\sigma_-{}^k$ for $a, b \in R$, and $\partial\alpha = a\partial\sigma_+{}^k + b\partial\sigma_-{}^k = (a+(-1)^k b)\sigma_+{}^{k-1} + (b+(-1)^k a)\sigma_-{}^{k-1}$, so that $\alpha \in Z_k(\mathcal{S})$ if and only if $b = -(-1)^k a = (-1)^{k+1}a$, which implies $\alpha = a(\sigma_+{}^k + (-1)^{k+1}\sigma_-{}^k)$. If $k<n$, then $\alpha = a\partial\sigma_+{}^{k+1} = \partial(a\sigma_+{}^{k+1})$. Thus $H_k(\mathcal{S}) = 0$ for $1 \leq k \leq n-1$, and $\xi_n = \sigma_+{}^n + (-1)^{n+1}\sigma_-{}^n$ is a generator of $H_n(\mathcal{S}) \approx R$.

The antipodal map $T: \mathbf{S}^n \to \mathbf{S}^n$ is a regular cellular map with respect to the family of cells $\mathcal{S}$, and $T(\sigma_\pm{}^k) = \sigma_\mp{}^k$ for $0 \leq k \leq n$. Define $[T:\sigma_\pm{}^k:\sigma_\pm{}^k] = 0$ and $[T:\sigma_\pm{}^k:\sigma_\mp{}^k] = 1$ for $0 \leq k \leq n$. Clearly T is a proper cellular map which satisfies conditions (i) and (ii) of 4.7. If $k \geq 1$, then

$$[T:\sigma_\pm{}^k:\sigma_\mp{}^k][\sigma_\mp{}^k:\sigma_\pm{}^{k-1}] = 1\cdot(-1)^k = [\sigma_\pm{}^k:\sigma_\mp{}^{k-1}][T:\sigma_\mp{}^{k-1}:\sigma_\pm{}^{k-1}],$$

and

$$[T:\sigma_\pm{}^k:\sigma_\mp{}^k][\sigma_\mp{}^k:\sigma_\mp{}^{k-1}] = 1\cdot 1 = [\sigma_\pm{}^k:\sigma_\pm{}^{k-1}][T:\sigma_\pm{}^{k-1}:\sigma_\mp{}^{k-1}],$$

so that (iii) is satisfied as well. Now note that $T_\#: C_*(\mathcal{S}) \to C_*(\mathcal{S})$ is given by the formula $T_\#(a\sigma_+{}^k + b\sigma_-{}^k) = a\sigma_-{}^k + b\sigma_+{}^k$, and, in particular, $T_\#(\xi_n) = T_\#(\sigma_+{}^n + (-1)^{n+1}\sigma_-{}^n) = \sigma_-{}^n + (-1)^{n+1}\sigma_+{}^n = (-1)^{n+1}\xi_n$. Thus in homology, $T_*(\xi_n) = (-1)^{n+1}\xi_n$. Making use of the natural transformation Ψ from cellular to singular homology, we have proved the following.

THEOREM. *If $T: \mathbf{S}^n \to \mathbf{S}^n$ is the antipodal map and $\xi_n \in H_n(\mathbf{S}^n)$, then $T_*(\xi_n) = (-1)^{n+1}\xi_n$.* ▮

THEOREM **4.11.** *The map $\Psi\lambda_*$ of 4.5 is a natural equivalence of $H_*'(K)$ with $H_*(|K|)$ for any abstract simplicial complex K.*

Proof. Let $f: K \to L$ be a simplicial map and let $\mathcal{S}$ and $\mathcal{J}$ be the cell structures on $|K|$ and $|L|$, respectively. Then $|f|$ has a contractible and therefore a proper carrier. By 4.7, if for any orientations of $|K|$ and $|L|$ we can find a chain map $f'':C_*(\mathcal{S})\to C_*(\mathfrak{J})$ for which $f''(\sigma)$ is a linear combination of the faces of $f(\sigma)$ with unit coefficients and for which, for each vertex v_0, we have $f''(v_0) = |f|(v_0)$, then f'' is $|f|_\#$, the chain map induced by $|f|$. Let $f':C'_*(K)\to C'_*(L)$ be the usual induced map of simplicial chains which sends $\langle v_0, \cdots, v_n\rangle$ into $\langle f(v_0), \cdots, f(v_n)\rangle$ if the latter is nondegenerate and 0 otherwise. Then $\lambda_L f'\lambda_K^{-1} = f''$ satisfies the conditions above and therefore is $|f|_\#$. Thus λ_K is a chain isomosphism natural with respect to order preserving maps. Since Ψ is natural with respect to order preserving maps, it follows that $\Psi\lambda_*$ is an isomorphism of $H'_*(K)$ with $H_*(|K|)$ natural with respect to simplicial maps which preserve some ordering on the vertices. But since $\Psi\lambda_*$ is independent of the vertex ordering chosen and since we may find compatible orderings for any simplicial map, it follows that $\Psi\lambda_*$ is natural with respect to all simplicial maps. ∎

Example **4.12.** Let $(X,\mathcal{S})$ and $(Y,\mathfrak{J})$ be oriented regular CW complexes. Let $f:X\otimes Y\to Y\otimes X$ be the map defined by $f(x,y) = (y,x)$. Then for $\sigma\times\tau\in\mathcal{S}\otimes\tau$, $f(\sigma\times\tau) = \tau\times\sigma$, and f is clearly a regular cellular map.

If we give $(X\otimes Y,\ \mathcal{S}\otimes\mathfrak{J})$ and $(Y\otimes X,\ \mathfrak{J}\otimes\mathcal{S})$ the orientations described in theorem 4.3, then $f_\#:C_*(\mathcal{S}\otimes\mathfrak{J})\to C_*(\mathfrak{J}\otimes\mathcal{S})$ has the property

$$f_\#(\sigma^p\times\tau^q) = (-1)^{pq}\tau^q\times\sigma^p.$$

By corollary 4.9, f is a proper map. Consider the elements $b^{p,q}_{\lambda,\mu,\lambda',\mu'} = (-1)^{pq}\delta_{\lambda,\lambda'}\delta_{\mu,\mu'}\in R$, where the δ's are Kronecker deltas. Then

(i) if $\tau^r_{\mu'}\times\sigma^{p+q-r}_{\lambda'}$ is not a subset of $f(\sigma_\lambda{}^p\times\tau_\mu{}^q)$, $b^{p,q}_{\lambda,\mu,\lambda',\mu'} = 0$;
(ii) $b^{0,0}_{\lambda,\mu,\lambda',\mu'} = 1$;
(iii) if $p+q\geq 1$, then $(-1)^{pq}[\tau^q\times\sigma^p:\tau^{q-1}\times\sigma^p]$
$+(-1)^{pq}[\tau^q\times\sigma^p:\tau^q\times\sigma^{p-1}]$

$$=(-1)^{pq}[\tau^q:\tau^{q-1}]+(-1)^{pq}(-1)^q[\sigma^p:\sigma^{p-1}]$$

$$=(-1)^{p(q-1)}[\sigma^p\times\tau^q:\sigma^p\times\tau^{q-1}]+(-1)^{q(p-1)}[\sigma^p\times\tau^q:\sigma^{p-1}\times\tau^q].$$

By theorem 4.7, $f_\#(\sigma^p\times\tau^q) = (-1)^{pq}\tau^q\times\sigma^p$.

5. *QUOTIENT COMPLEXES*

Let X be a CW complex and $\mathcal{R}$ a cellular equivalence relation such that $X/\mathcal{R}$ is a CW complex, and let $p:X\to X/\mathcal{R}$ be the regular cellular quotient map. Let $\hat{\mathcal{S}}$ be a set consisting of one minimal cell from each $\mathcal{R}$-equivalence class, and call this a set of representative minimal cells. For any such $\hat{\mathcal{S}}$, there is a bijective correspondence with the cells of a quotient structure on $X/\mathcal{R}$. If σ is a cell of X, let $\hat{\sigma}$ denote the equivalent representative cell in $\hat{\mathcal{S}}$. An orientation of the cells of $\hat{\mathcal{S}}$, in particular an orientation of X, defines an orientation of $X/\mathcal{R}$, called the *quotient orientation.* From the definition of cellular equivalence relation, given any two equivalent minimal n-cells σ,σ' of X, there is a homeomorphism $h_{\sigma',\sigma}:E^n\to E^n$ such that $p\varphi_\sigma = p\varphi_{\sigma'}h_{\sigma',\sigma}$. If X is oriented, then $h_{\sigma',\sigma}$ is a map of oriented n-cells, and we let $\epsilon(\sigma',\sigma)$ be its degree. If $\epsilon(\sigma',\sigma)$ is a unit of R other than 1, the selection of σ' from its equivalence class rather than σ for the representative cell $\hat{\sigma}$ would change the quotient orientation on $X/\mathcal{R}$. If X is so oriented that $\epsilon(\sigma',\sigma)=1$ for any two equivalent minimal cells, we say that X is *$\mathcal{R}$-oriented.*

THEOREM **5.1.** *Let $(X,\mathcal{S})$ be an oriented CW complex, let $\mathcal{R}$ be a cellular equivalence relation such that $X/\mathcal{R}$ is a CW complex, and let $p:X\to X/\mathcal{R}$ be the quotient map. Let $X/\mathcal{R}$ have the quotient orientation with respect to some set $\hat{\mathcal{S}}$ of representative minimal cells. Then $p_\#:C_*(\mathcal{S})\to C_*(\mathcal{S}/\mathcal{R})$ is an epimorphism whose kernel $Q(\mathcal{R})$ is the chain submodule generated by all σ which are not $\mathcal{R}$-minimal and all $\hat{\sigma}-\epsilon(\hat{\sigma},\sigma)\sigma$, where σ is minimal and $\hat{\sigma}$ is its equivalent in $\hat{\mathcal{S}}$. In particular, if X is $\mathcal{R}$-oriented, then $Q(\mathcal{R})$ is generated by all σ which are not minimal and all $\hat{\sigma}-\sigma$, where σ is minimal.*

Proof. The map $p_\#$ is a map of free R-modules sending each nonminimal basic generator σ into 0 (because p maps the cell σ into a cell of lower dimension) and the minimal generator σ into $\epsilon(\sigma,\hat{\sigma})p_\#(\hat{\sigma})$. Hence $Q(\mathcal{R})\subset \operatorname{Ker} p_\#$. Since a subset of $\{p_\#(\sigma)\mid\sigma\in\hat{\mathcal{S}}\}$ is independent if the corresponding set of representative cells consists of distinct elements, $\operatorname{Ker} p_\#$ cannot be larger than $Q(\mathcal{R})$. Thus $\operatorname{Ker} p_\# = Q(\mathcal{R})$. ∎

Similarly, we get the following result.

THEOREM 5.2. *Let $f:X\to X'$ be a regular cellular map of CW complexes, let $\mathcal{R},\mathcal{R}'$ be cellular equivalence relations for which $X/\mathcal{R}$, $X'/\mathcal{R}'$ are CW complexes, let f be $\mathcal{R}$-$\mathcal{R}'$-compatible, and let $\bar{f}:X/\mathcal{R}\to X'/\mathcal{R}'$ be the induced map on the quotient spaces. Then the following diagram is commutative.*

$$\begin{array}{ccccccccc} 0 \to & Q(\mathcal{R}) & \to & C_*(\mathcal{S}) & \to & C_*(\mathcal{S}/\mathcal{R}) & \to 0 \\ & \downarrow f_\# & & \downarrow f_\# & & \downarrow \bar{f}_\# & \\ 0 \to & Q(\mathcal{R}') & \to & C_*(\mathcal{S}') & \to & C_*(\mathcal{S}'/\mathcal{R}') & \to 0 \end{array} .$$

In particular, if X and X' are $\mathcal{R},\mathcal{R}'$-oriented, then for any $\mathcal{R}$-minimal cell σ of X, we have $\bar{f}_\#(\bar{\sigma})=0$ if $f(\sigma)$ is not $\mathcal{R}'$-minimal in X', and $\bar{f}_\#(\bar{\sigma})=\overline{f_\#(\sigma)}$ if $f(\sigma)$ is $\mathcal{R}'$-minimal in X'. ∎

We now apply these results in the case that $\mathcal{R}$ is the relation induced by a family of identifications on a CW complex X. This device is especially useful for homological calculations, and we work out several examples in detail. We refer the reader to I.6.9 and II.5.8 for the proof that if Ω is a family of identifications on a CW complex $(X,\mathcal{S})$, then the quotient complex $(X/\Omega,\ \mathcal{S}/\Omega)$ is a CW complex, the identification map $p:X\to X/\Omega$ is a regular cellular map, and that $(X/\Omega,\ \mathcal{S}/\Omega)$ is a normal CW complex whenever $(X,\mathcal{S})$ is.

If $\mathcal{R}_\Omega$ is the equivalence relation induced by the family of identifications Ω, then every cell is $\mathcal{R}_\Omega$-minimal, since the identifications $\omega\in\Omega$ are regular cellular homeomorphisms. Commutativity of the diagram

$$\begin{array}{ccc} C_n(\mathcal{S}) & \xrightarrow{\partial} & C_{n-1}(\mathcal{S}) \\ \downarrow p_\# & & \downarrow p_\# \\ C_n(\mathcal{S}/\Omega) & \xrightarrow{\partial_\Omega} & C_{n-1}(\mathcal{S}/\Omega) \end{array}$$

and knowledge of the degrees $[\omega:\sigma:\sigma']$, where $\omega(\sigma)=\sigma'$ enables us to calculate $\partial_\Omega(\bar{\sigma}) = p_\# \partial(\hat{\sigma})$ for any cell $\hat{\sigma}$ in the class of $\bar{\sigma}$. Note that if $(X, \mathcal{S})$ is a regular CW complex, then by 4.2 and 4.7, incidence numbers of X and degrees of $\omega \in \Omega$ can be determined from the incidence and degree relations, and the calculation of ∂_Ω may be carried out. The following three examples illustrate this method of calculation.

Example **5.3.** Real projective space $\mathbf{RP}^n$. Let the n-sphere $\mathbf{S}^n$ have the regular cellular decomposition of I.2.2 into cells $\{\sigma_+{}^0, \sigma_-{}^0, \cdots, \sigma_+{}^n, \sigma_-{}^n\}$. If $T: \mathbf{S}^n \to \mathbf{S}^n$ is the antipodal map, T is a regular cellular homeomorphism, and $T(\sigma_\pm{}^k) = \sigma_\mp{}^k$. Clearly $\Omega = \{T, \text{identity}\}$ is a family of identifications on $\mathbf{S}^n$, and $\mathbf{S}^n/\Omega = \mathbf{RP}^n$. Let $p: \mathbf{S}^n \to \mathbf{RP}^n$ be the identification map. The complex $\mathbf{RP}^n$ has cells $\{\tau^0, \cdots, \tau^n\}$.

Let the cells $\sigma_\pm{}^k$ of $\mathbf{S}^n$ have the orientation given in 4.10, and give τ^k the orientation $\tau^k = p_\#(\sigma_+{}^k)$. Then we see that $\partial_\Omega(\tau^k) = p_\#\partial(\sigma_+{}^k) = p_\#(\sigma_+{}^{k-1} + (-1)^k \sigma_-{}^{k-1}) = \tau^{k-1} + (-1)^k p_\#(\sigma_-{}^{k-1})$. But $p_\#(\sigma_-{}^{k-1}) = p_\# T_\#(\sigma_+{}^{k-1}) = p_\#(\sigma_+{}^{k-1}) = \tau^{k-1}$. Thus $\partial_\Omega(\tau^k) = (1 + (-1)^k)\tau^{k-1}$ for $k \geq 1$. If $\mathfrak{I} = \{\tau^0, \cdots, \tau^n\}$, then $C_k(\mathfrak{I}) = R\tau^k$ for $0 \leq k \leq n$. Thus,

$$Z_{2k-1}(\mathfrak{I}) = R\tau^{2k-1} \quad \text{and} \quad B_{2k-1}(\mathfrak{I}) = 2R\tau^{2k-1} \quad \text{for} \quad 2k-1 < n,$$

$$Z_{2k}(\mathfrak{I}) = {}^2R\tau^{2k} \quad \text{and} \quad B_{2k}(\mathfrak{I}) = 0 \quad \text{for} \quad 0 < 2k \leq n,$$

where for each integer n, ${}^nR = \{r \in R \mid nr = 0\}$. Passing to homology, we obtain the following.

(i) If n is even:

$$H_0(\mathbf{RP}^n) \approx R;$$

$$H_{2k-1}(\mathbf{RP}^n) \approx R/2R \quad \text{for} \quad 2k-1 < n;$$

$$H_{2k}(\mathbf{RP}^n) \approx {}^2R \quad \text{for} \quad 0 < 2k \leq n;$$

$$H_k(\mathbf{RP}^n) = 0 \quad \text{for} \quad n < k.$$

(ii) If n is odd:

$$H_0(\mathbf{RP}^n) \approx R;$$

$$H_{2k-1}(\mathbf{RP}^n) \approx R/2R \quad \text{for} \quad 2k-1<n;$$

$$H_{2k}(\mathbf{RP}^n) \approx {}^2R \quad \text{for} \quad 0<2k<n;$$

$$H_n(\mathbf{RP}^n) \approx R;$$

$$H_k(\mathbf{RP}^n) = 0 \quad \text{for} \quad n<k.$$

Example **5.4.** The lens spaces $\mathbf{L}^n(p,q)$ and $\mathbf{L}^\infty(p,1)$. Recall that in the definition of the lens spaces I.2.6, we defined a regular cell complex on $\mathbf{S}^{2n-1}$ which had p cells in each dimension, and that a cell structure $(\mathbf{L}^n(p,q),\mathfrak{J})$ was obtained by identifying cells via a certain rotation T. The space $\mathbf{L}^n(p,q)$ may be regarded as obtained from a family of identifications Ω consisting of the restrictions of T and its iterates to the cells of $\mathfrak{S}$. Proceeding in a manner analogous to that used in the calculations for $\mathbf{RP}^n$, we find that $C_m(\mathfrak{J}) \approx R$ for $0 \leq m \leq 2n-1$, and that $\partial(\tau^{2r-1}) = 0$, $\partial(\tau^{2r}) = p\tau^{2r-1}$, where τ^m is the generator of $C_m(\mathfrak{J})$. Passing to homology, we obtain

$$H_0(\mathbf{L}^n(p,q)) \approx R;$$

$$H_{2r-1}(\mathbf{L}^n(p,q)) \approx R/pR \quad \text{for} \quad 1 \leq r \leq n-1;$$

$$H_{2r}(\mathbf{L}^n(p,q)) \approx {}^pR \quad \text{for} \quad 1 \leq r \leq n-1;$$

$$H_{2n-1}(\mathbf{L}^n(p,q)) \approx R.$$

A similar calculation yields

$$H_0(\mathbf{L}^\infty(p,1)) \approx R;$$

$$H_{2r-1}(\mathbf{L}^\infty(p,1)) \approx R/pR \quad \text{for} \quad 1 \leq r;$$

$$H_{2r}(\mathbf{L}^\infty(p,1)) \approx {}^pR \quad \text{for} \quad 1 \leq r.$$

Example **5.5.** Compact 2-manifolds. Let $\sigma^2 = \{z \in \mathbf{C} \mid |z| \leq 1\}$, and give σ^2 a regular CW structure with 0-cells $\sigma_k^0 = e^{\pi i(k-1)/n}$ for $1 \le k \le 2n$; 1-cells $\sigma_k^1 = \{e^{\pi i(2k-1+t)/2n} \mid -1 \le t \le 1\}$ for $1 \le k \le 2n$; and a 2-cell σ^2. The characteristic maps for the 0-cells are the obvious ones; those for the 1-cells are the maps $\varphi_k^1: D^1 \to \sigma_k^1 \subset \sigma^2$ defined by $\varphi_k^1(t) = e^{\pi i(2k-1+t)/2n}$, and the map for the 2-cell is the identity map. Note that $\dot{\sigma}_k^1 = \sigma_k^0 \cup \sigma_{k-1}^0$ (where $\sigma_0^0 = \sigma_{2n}^0$) and $\dot{\sigma}^2 = \mathbf{U}_k\, \sigma_k^1$.

By theorem 4.2, we may orient the CW complex $(\sigma^2, \mathcal{S})$ by choosing:

$$[\sigma_k^1 : \sigma_k^0] = 1 = -[\sigma_k^1 : \sigma_{k-1}^0] \quad \text{for} \quad 1 \le k \le 2n,$$

$$[\sigma^2 : \sigma_k^1] = 1 \qquad \text{for} \quad 1 \le k \le 2n.$$

Case 1. *Orientable 2-manifolds.* Let n be even, say $n = 2m$, and let Ω_n be the smallest set of identifications on σ^2 containing the homeomorphisms $\omega_k: \sigma_k^1 \to \sigma_{k+n}^1$ defined by $\omega_k(z) = \bar{z}e^{\pi i(n+2k-1)/n}$ for $1 \le k \le n$. Observe that the diagram

$$\begin{array}{ccc} (D^1, \dot{D}^1) & \xrightarrow{\varphi_k^1} & (\sigma_k^1, \dot{\sigma}_k^1) \\ \downarrow{\scriptstyle \iota} & & \downarrow{\scriptstyle \omega_k} \\ (D^1, \dot{D}^1) & \xrightarrow{\varphi_{n+k}^1} & (\sigma_{k+n}^1, \dot{\sigma}_{k+n}^1) \end{array}$$

commutes, where $\iota: D^1 \to D^1$ is the map $\iota(t) = -t$. This implies that $\omega_{k*}(\sigma_k^1) = -\sigma_{k+n}^1$ for the algebraic cell σ_k^1.

Let $p: \sigma^2 \to \sigma^2/\Omega_n = \mathbf{X}_n$ be the quotient map. Then $\mathbf{X}_n$ has a single 0-cell $\tau^0 = p(\sigma_k^0)$ for $1 \le k \le 2n$; $n = 2m$ 1-cells $\tau_k^1 = p(\sigma_k^1) = p(\sigma_{k+n}^1)$ for $1 \le k \le n$; and a single 2-cell $\tau^2 = p(\sigma^2)$. The cells $\mathfrak{J}$ of $\mathbf{X}_n$ are oriented as follows:

$$\tau^0 = p_{\#}(\sigma_k^0) \qquad \text{for} \quad 1 \le k \le 2n;$$

$$\tau_k^1 = p_{\#}(\sigma_k^1)$$

$$= -p_{\#}(\sigma_{k+n}^1) \quad \text{for} \quad 1 \le k \le n;$$

$$\tau^2 = p_{\#}(\sigma^2).$$

Since each 1-cell τ_k^1 is incident with the single 0-cell τ^0, we have $\partial_\Omega\tau_k^1=0$ by 3.11 (ii). Thus $Z_1(\mathfrak{I})=C_1(\mathfrak{I})\approx R^n$. Also, we see that

$$\partial_\Omega\tau^2=\partial_\Omega p_\#(\sigma^2)=p_\#\partial_\Omega(\sigma^2)=p_\#(\sum_{k=1}^{2n}\sigma_k^1)$$

$$=\sum_{k=1}^{n}(p_\#(\sigma_k^1)+p_\#(\sigma_{k+n}^1))=0.$$

Thus $B_1(\mathfrak{I})=0$, and $C_2(\mathfrak{I})=Z_2(\mathfrak{I})\approx R$. We conclude that

$$H_0(\mathbf{X}_n)\approx R;$$

$$H_1(\mathbf{X}_n)\approx R^n;$$

$$H_2(\mathbf{X}_n)\approx R;$$

$$H_q(\mathbf{X}_n)=0 \qquad \text{for} \quad q>2.$$

Case 2. Nonorientable 2-manifolds. Let n be an arbitrary positive integer, and let Ω_n be the smallest set of identifications on σ^2 containing the homeomorphisms $\omega_k:\sigma_k^1\to\sigma_{k+n}^1$ defined by $\omega_k(z)=\bar{z}e^{\pi i(n+2k-1)/n}$ for $1\leq k\leq n-1$, and $\omega_n(z)=ze^{\pi i}$. One easily sees that for the algebraic cells σ_k^1, we have $\omega_{k*}(\sigma_k^1)=-\sigma_{k+n}^1$ for $1\leq k\leq n-1$, and $\omega_{n*}(\sigma_n^1)=\sigma_{2n}^1$.

Let $p:\sigma^2\to\sigma^2/\Omega_n=\mathbf{Y}_n$ be the quotient map. Then $\mathbf{Y}_n$ has a single 0-cell $\tau^0=p(\sigma_k^0)$ for $1\leq k\leq 2n$; n 1-cells $\tau_k^1=p(\sigma_k^1)=p(\sigma_{k+n}^1)$ for $1\leq k\leq n$; and a single 2-cell $\tau^2=p(\sigma^2)$. The cells $\mathfrak{I}$ of $\mathbf{Y}_n$ are oriented as follows:

$$\tau^0=p_\#(\sigma_k^0) \qquad \text{for} \quad 1\leq k\leq 2n;$$

$$\tau_k^1=p_\#(\sigma_k^1)=-p_\#(\sigma_{n+k}^1) \quad \text{for} \quad 1\leq k\leq n-1;$$

$$\tau_n^1=p_\#(\sigma_n^1)=p_\#(\sigma_{2n}^1);$$

$$\tau^2=p_\#(\sigma^2).$$

The calculation of homology is the same as in the orientable case, except that $\partial_\Omega \tau^2 = 2\tau_n^1$. We conclude that

$$H_0(\mathbf{Y}_n) \approx R;$$

$$H_1(\mathbf{Y}_n) \approx R^{n-1} \oplus R/2R;$$

$$H_2(\mathbf{Y}_n) \approx {}^2R;$$

$$H_q(\mathbf{Y}_n) = 0 \qquad \text{for} \quad q > 2.$$

We remark that any compact 2-manifold without boundary except the 2-sphere can be obtained from a family of identifications on σ^2 as above. See Massey [23].

An analysis similar to that given above enables us to calculate the degrees of an induced map $\bar{f}: X/\Omega \to X'/\Omega'$, where $f: X \to X'$ is a regular cellular map compatible with families of identifications Ω on X and Ω' on X'. We omit the details.

Another example of theorem 5.1 is the description of the cellular chains of the geometric realization $|X|$ of an ssc X. This example will be worked out later in section 7.

As a final example, let X be the union of two subcomplexes X_1 and X_2, and let $p: X_1 \cup X_2 \to X$ map the disjoint union onto X by the inclusion map on each summand. Let $\mathcal{R}$ be the cellular relation identifying the two copies of $X_1 \cap X_2$ in the two summands. Then the exact homology sequence of the short exact sequence of chain modules

$$0 \to Q(\mathcal{R}) \to C_*(\mathcal{S}_1) \oplus C_*(\mathcal{S}_2) \to C_*(\mathcal{S}) \to 0$$

is the Mayer–Vietoris sequence of the excisive decomposition of X into $\{X_1, X_2\}$ (see [35]).

6. *PRODUCT AND ADJUNCTION COMPLEXES*

The close relation between the cellular chains and the (geometric) cellular structure on a CW complex leads to a close parallel between (geometric) cellular constructions with CW complexes and some corresponding constructions on the cellular chains. The last section illustrated this for quotient complexes, and this section concerns the other constructions we have developed.

By II.5.3, the spaces $X \otimes Y$ and $X \times Y$ have the same compact subsets for any CW complexes X and Y. Thus it follows from axiom C for singular homology that the natural bijection $\eta: X \otimes Y \to X \times Y$ induces isomorphisms of homology in all dimensions. For this reason, we can drop the $\otimes$ notation on the homology level.

In dealing with product complexes, we will assume the following facts which may be found in Spanier [35].

(1) If (X,A) and (Y,B) are pairs such that $\{A \times Y, X \times B\}$ is an excisive couple, there is a natural transformation

$$T: H_*(X,A) \otimes H_*(Y,B) \to H_*(X \times Y, (A \times Y) \cup (X \times B))$$

which is an isomorphism if either $H_*(X,A)$ or $H_*(Y,B)$ consists of free R-modules. Such a transformation T is called an Eilenberg–Zilber transformation. We will denote $T(x \otimes y)$ by $x \times y$.

(2) If $\{(X_1,A_1),(X_2,A_2)\}$ is an excisive couple of pairs in X, then under the boundary operators of the appropriate Meyer–Vietoris sequences

$$\begin{array}{ccc}
H_{p+q}((X_1 \cup X_2, A_1 \cup A_2) \times (Y,B)) & \xrightarrow{\Delta} & H_{p+q-1}((X \cap X_2, A_1 \cap A_2) \times (Y,B)) \\
\uparrow T & & \uparrow T' \\
H_p(X_1 \cup X_2, A_1 \cup A_2) \otimes H_q(Y,B) & \xrightarrow{\Delta' \otimes \mathrm{id}} & H_{p-1}(X_1 \cap X_2, A_1 \cap A_2) \otimes H_q(Y,B)
\end{array}$$

$$\begin{array}{ccc}
H_{p+q}((Y,B) \times (X_1 \cup X_2, A_1 \cup A_2)) & \xrightarrow{\Delta} & H_{p+q-1}((Y,B) \times (X_1 \cap X_2, A_1 \cap A_2)) \\
\uparrow T & & \uparrow T' \\
H_p(Y,B) \otimes H_q(X_1 \cup X_2, A_1 \cup A_2) & \xrightarrow{(-1)^p \mathrm{id} \otimes \Delta'} & H_p(Y,B) \otimes H_{p+q-1}(X_1 \cap X_2, A_1 \cap A_2)
\end{array}$$

are commutative diagrams. See Spanier [35, p. 250, Theorem 6]. Of course, in order that the appropriate Meyer–Vietoris sequences exist, $\{(X_1,A_1),(X_2,A_2)\}$ and $\{(Y,B) \times (X_1,A_1), (Y,B) \times (X_2,A_2)\}$ must be excisive couples of pairs.

As a special case of this we obtain the following.

Proposition 6.1. *If (X,C,D) is a CW triple and (Y,B) is a CW pair, the following diagrams are commutative*

$$\begin{array}{ccc} H_{p+q}((Y,B)\times(X,C)) & \xrightarrow{\Delta} & H_{p+q-1}((Y,B)\times(C,D)) \\ \uparrow T & & \uparrow T' \\ H_p(Y,B)\otimes H_q(X,C) & \xrightarrow{(-1)^p \mathrm{id}\otimes\Delta'} & H_p(Y,B)\otimes H_{q-1}(C,D) \end{array}$$

$$\begin{array}{ccc} H_{p+q}((X,C)\times(Y,B)) & \xrightarrow{\Delta} & H_{p+q-1}((C,D)\times(Y,B)) \\ \uparrow T & & \uparrow T' \\ H_p(X,C)\otimes H_q(Y,B) & \xrightarrow{\Delta\otimes \mathrm{id}} & H_{p-1}(C,D)\otimes H_q(Y,B). \end{array}$$

Proof. Since pairs of CW complexes are excisive, the appropriate Meyer–Vietoris sequences exist if we take $X=X_1$, $D=A_1$, and $X_2=A_2=C$ in the diagrams above. ▮

We also need the following fact about the boundary operator of a triple.

Lemma 6.2. *Let X be a CW complex with subcomplexes A, B, C such that $X\supseteq A\cup B\supseteq C\supseteq A\cap B$. Let*

$$\Delta: H_n(X,A\cup B)\to H_{n-1}(A\cup B,C)$$

$$\Delta_B: H_n(X,A\cup B)\to H_{n-1}(A\cup B,A\cup(B\cap C))\approx H_{n-1}(B,B\cap C)$$

$$\Delta_A: H_n(X,A\cup B)\to H_{n-1}(A\cup B,B\cup(A\cap C))\approx H_{n-1}(A,A\cap C)$$

be boundary operators of appropriate triples. Let

$$k_B: H(A\cup B,A\cup(B\cap C))\xrightarrow{\approx} H(B,B\cap C)\xrightarrow{i_B} H(A\cup B,C)$$

$$k_A: H(A\cup B,B\cup(A\cap C))\xrightarrow{\approx} H(A,A\cap C)\xrightarrow{i_A} H(A\cup B,C)$$

be compositions of excisions and maps induced by inclusions.

Then (i) k_A, k_B *are injective and* $\operatorname{Im} k_A \oplus \operatorname{Im} k_B \approx H(A \cup B, C)$, *and* (ii) $\Delta = k_A \Delta_A + k_B \Delta_B$.

Proof. Note that i_A, i_B correspond to maps $\bar{\imath}_A, \bar{\imath}_B$ mapping $B/(B \cap C)$, $A/(A \cap C)$ into $(A \cup B)/C$, and since $C \supset A \cap B$, $(A \cup B)/C = \operatorname{Im} i_A \vee \operatorname{Im} i_B$. Thus we can apply 6.1, proving (i).

Consider the commutative diagram:

$$\begin{array}{ccc} H(A \cup B) & \xrightarrow{(j_B)_*} & H(B \cup A, A \cup (B \cap C)) \\ \Big\downarrow (j_A)_* & \searrow j_* & \Big\downarrow k_B \\ H(A \cup B, B \cup (A \cap C)) & \xrightarrow{k_A} & H(A \cup B, C) \end{array}$$

where j, j_A, and j_B are inclusion maps.

Let p_A, p_B mapping $(A \cup B)/C$ onto $A/(A \cap C)$, $B/(B \cap C)$ be the projections of the wedge sum onto its summands. Thus $p_A \bar{\imath}_A = 1_A, p_B \bar{\imath}_B = 1_B$, and $p_A \bar{\imath}_B, p_A \bar{\imath}_B$ are point maps, and these maps induce homomorphisms which give $H_*((A \cup B)/C)$ the structure of direct sum $(i_A)_* H(A/(A \cap C)) \oplus (i_B)_* H(B/(B \cup C))$. In particular, $(\bar{\imath}_A)_*(p_A)_* + (\bar{\imath}_B)_*(p_B)_* = 1$.

Let $\bar{\jmath}_A : A \cup B \to (A \cup B)/(B \cup (A \cap C)) \to A/(A \cap C)$

$\bar{\jmath}_B : A \cup B \to (A \cup B)/(A \cup (B \cap C)) \to B/(B \cap C)$

correspond to j_A, j_B. Then $p_A \bar{\jmath} = \bar{\jmath}_A$, $p_B \bar{\jmath} = \bar{\jmath}_B$.

$$\begin{aligned} \text{Thus } k_A(j_A)_* + k_B(j_B)_* &= (\bar{\imath}_A)_*(p_A \bar{\jmath})_* + (\bar{\imath}_B)_*(p_B \bar{\jmath})_* \\ &= [(\bar{\imath}_A)_*(p_A)_* + (\bar{\imath}_B)_*(p_B)_*] \cdot \bar{\jmath}_* \\ &= \bar{\jmath}_*. \end{aligned}$$

Since Δ is by definition the composition

$$H_n(X, A \cup B) \xrightarrow{\partial} H_{n-1}(A \cup B) \xrightarrow{j_*} H_{n-1}(A \cup B, C)$$

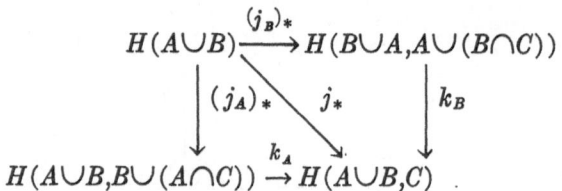

we have:

$$j_*\partial = \jmath_*\cdot\partial = k_A(\jmath_A)_*\partial + k_B(\jmath_B)_*\partial = k_A\Delta_A + k_B\Delta_B,$$

since also by definition, $\Delta_A = (j_A)_*\partial$, and $\Delta_B = (j_B)\partial$. ∎

Definition **6.3.** If $C_* = \{C_q, \partial_q\}$ and $C'_* = \{C_q', \partial_q'\}$ are R-chain modules, their tensor product $(C\otimes C')_*$ is the graded chain module with $(C\otimes C')_q = \sum_i \oplus C_i \otimes C'_{q-i}$ and $\partial_q(c_i\otimes c'_{q-i}) = \partial_i(c_i)\otimes c'_{q-i} + (-1)^i c_i \otimes \partial'_{q-i}(c'_{q-i})$, where $c_i \in C_i$ and $c'_{q-i}\in C'_{q-i}$.

THEOREM **6.4.** *Let* $(X,\mathcal{S})$ *and* $(Y,\mathfrak{T})$ *be CW complexes. There is a natural chain equivalence* $(C(\mathcal{S})\otimes C(\mathfrak{T}))_* \to C_*(\mathcal{S}\otimes\mathfrak{T})$.

Proof. Let $W = X^n\times Y^m$, $A = X^n \times Y^{m-1}$, $B = X^{n-1}\times Y^m$, and $C = (X^n\times Y^{n-2})\cup(X^{n-1}\times Y^{m-1})\cup(X^{n-2}\times Y^m)$, so that $W\supset A\cup B\supset C\supset A\cap B$. Then lemma 6.2 applies and we have $\Delta = k_A\Delta_A + k_B\Delta_B$. By proposition 1.3, the map $k_A: H_*(A\cup B, B\cup(A\cap C))\to H_*(A\cup B, C)$ is equivalent to the map induced by the inclusion $A/(A\cap C)\to (A/(A\cap C)\vee B/(B\cap C))$ and is therefore the injection which identifies the direct summand generated by the cells of A outside C with the corresponding generators of the cellular chains of $(A\cup B)/C$. Similarly k_B is the injection of the direct summand generated by the cells of B outside C into the cellular chains of $(A\cup B)/C$. The maps Δ_A and Δ_B are boundary operators in the exact sequences of triples. By proposition 6.1 there are Eilenberg–Zilber transformations T, T' such that the following diagram is commutative.

$$\begin{array}{ccc}
H_{n+m}((X^n,X^{n-1})\times(Y^m,Y^{m-1})) & \xrightarrow{\Delta_B} & H_{n+m-1}((X^n,X^{n-1})\times(Y^{m-1},Y^{m-2})) \\
\uparrow T & & \uparrow T' \\
H_n(X^n,X^{n-1})\otimes H_m(Y^m,Y^{m-1}) & \xrightarrow{(-1)^n \mathrm{id}\otimes\partial'} & H_n(X^n,X^{n-1})\otimes H_{m-1}(Y^{m-1},Y^{m-2}).
\end{array}$$

Thus we may identify Δ_B with $(-1)^n\,\mathrm{id}\otimes\partial'$, and by a similar diagram, we may identify Δ_A with $\partial\otimes\mathrm{id}$, so that if $\sigma^n\times\tau^m$ is a cell of $\mathcal{S}\otimes\mathfrak{T}$, then $\Delta(\sigma^n\times\tau^m) = \partial(\sigma^n)\times\tau^m + (-1)^n\sigma^n\times\partial'(\tau^m)$.

Hence if we let $T: \sum_i \oplus H_i(X^i, X^{i-1}) \otimes H_{p-i}(Y^{p-i}, Y^{p-i-1}) \to H_p((X \times Y)^p, (X \times Y)^{p-1})$ be a composition of Eilenberg–Zilber maps with maps induced by inclusions of $(X^i, X^{i-1}) \times (Y^{p-i}, Y^{p-i-1})$ in $((X \times Y)^p, (X \times Y)^{p-1})$, then T is a natural chain equivalence of $(C(\mathcal{S}) \otimes C(\mathcal{T}))_*$ with $C_*(\mathcal{S} \otimes \mathcal{T})$. ∎

We point out that this is compatible with the result we obtained for oriented regular CW complexes in theorem 4.3.

Example **6.5.** A product of projective spaces $\mathbf{P} = \mathbf{RP}^{a_1} \times \cdots \times \mathbf{RP}^{a_n}$. Let each $\mathbf{RP}^{a_i}$ be given the cell structure $\mathcal{S}_i$ discussed in example 5.3, and let $\mathbf{P}$ be given the product structure $\mathcal{S}$. Then $C_*(\mathcal{S}) = C_*(\mathcal{S}_1) \otimes C_*(S_2) \otimes \cdots \otimes C_*(\mathcal{S}_n)$. If R is a ring of characteristic 2, then the boundary operator in each $C_*(\mathcal{S}_i)$ is zero, and therefore also in $C_*(\mathcal{S})$. Thus $H_q(\mathbf{P})$ is a free R-module of rank ρ_q, where ρ_q is the number of solutions in nonnegative integers of the equation $i_1 + i_2 + \cdots + i_n = q$.

Similarly, it is easy to see that if $N = a_1 + a_2 + \cdots + a_n$ is the dimension of $\mathbf{P}$, then the highest dimensional homology for a general ring R is

$H_N(\mathbf{P}) \approx R$ if $a_1, \cdots, a_n$ are all odd or R has characteristic 2;
$H_N(\mathbf{P}) \approx {}_2R$ otherwise.

Example **6.6.** The real Grassman variety $\mathbf{RG}(n,m)$. In order to simplify the calculations and notation, we will assume that R is a ring of characteristic 2, and we will write $X = \mathbf{RG}(n,m)$. The reader is referred to example I.2.5 for a description of the cellular decomposition $\mathcal{S}$ of X that we will use.

If $\sigma^N \in \mathcal{S}$ is a cell of X with type numbers $(p_n, \cdots, p_1)$ where $\sum_k kp_k = N$, we recall that there is a characteristic map $\varphi: E^N \to \sigma^N \subset X$ which is defined by setting $E^N = D^{p_n} \times D^{p_n + p_{n-1}} \times \cdots \times D^{p_n + p_{n-1} + \cdots + p_1}$ and defining $\varphi(v^1, v^2, \cdots, v^n)$ to be the subspace of $\mathbf{R}^{n+m}$ spanned by the row vectors of a certain $n \times m$ matrix in reduced echelon form. It is evident from the formula of example I.2.5 that we may factor φ as

$$\varphi: E^N \xrightarrow{p} \mathbf{P} \xrightarrow{\varphi'} \sigma^N \subset X,$$

where $\mathbf{P}$ is the product space $\mathbf{P} = \mathbf{RP}^{p_n} \times \mathbf{RP}^{p_n+p_{n-1}} \times \cdots \times \mathbf{RP}^{p_n+p_{n-1}+\cdots+p_1}$, and the map $p: E^N = D^{p_n} \times D^{p_n+p_{n-1}} \times \cdots \times D^{p_n+p_{n-1}+\cdots+p_1} \to \mathbf{P}$ is the product of the maps $D^{p_n+p_{n-1}+\cdots+p_i} \to \mathbf{RP}^{p_n+p_{n-1}+\cdots+p_i}$ which identify antipodal boundary points. If we let $\dot{\mathbf{P}}$ be the $(N-1)$-skeleton of $\mathbf{P}$, then $p(\dot{E}^N) \subset \dot{\mathbf{P}}$, and φ' induces $\varphi'': (\mathbf{P},\dot{\mathbf{P}}) \to (\sigma^N, \dot{\sigma}^N)$. Now consider the commutative diagram

$$
\begin{array}{ccccc}
 & & H_N(E^N,\dot{E}^N) & & \\
 & & \downarrow p_* & & \\
0 \to H_N(\mathbf{P}) & \xrightarrow{j'_*} & H_N(\mathbf{P},\dot{\mathbf{P}}) & \xrightarrow{\partial'_*} & H_{N-1}(\dot{\mathbf{P}}) \\
\downarrow \varphi''_* & & \downarrow \varphi''_* & & \downarrow \\
H_N(X^N) & \xrightarrow{j_*} & H_N(X^N,X^{N-1}) & \xrightarrow{\partial_*} & H_{N-1}(X^{N-1})
\end{array}
$$

where the horizontal lines are portions of the exact sequences of pairs.

From example 6.5, every cellular n-chain of $\mathbf{P}$ is a cycle (since R is of characteristic 2), and it follows that the map $j'_*: H_N(\mathbf{P}) \to H_N(\mathbf{P},\dot{\mathbf{P}})$ is an isomorphism of free R-modules of rank 1.

Let ζ_N be a generator of $H_N(E^N,\dot{E}^N)$ and let $\sigma^N \in C_N(\mathcal{S}) = H_N(X^N,X^{N-1})$ be an orientation of the N-cell σ^N. Then for some unit $u \in R$, we have $\sigma^N = \varphi_*(u\zeta_N) = \varphi''_* p_*(u\zeta_N) = \varphi''_* j'_*(\tau^N) = j_*\varphi'_*(\tau^N)$ for a suitable generator $\tau^N \in H_N(\mathbf{P})$. It follows that $\partial_*(\sigma^N) = \partial_* j_*\varphi'_*(\tau^N) = 0$, and since $\partial_N: C_N(\mathcal{S}) \to C_{N-1}(\mathcal{S})$ is the composition

$$
H_N(X^N,X^{N-1}) \xrightarrow{\partial_*} H_{N-1}(X^{N-1}) \xrightarrow{j_*} H_{N-1}(X^{N-1},X^{N-2}),
$$

we have $\partial_N(\sigma^N) = 0$. Since the integer N and the cell σ^N were chosen arbitrarily, we conclude that the chain complex $C_*(\mathcal{S}) = \{C_n(\mathcal{S}),\partial_n\}$ has a trivial boundary operator. From this we conclude the following.

If R is a ring of characteristic 2, then $H_q(\mathbf{RG}(n,m))$ is a free R-module of rank $\rho_q(n,m)$, where

$$
\rho_q(n,m) = \text{card}\,\{(p_n, \cdots, p_1) \in Z^n \mid p_k \geq 0 \quad \text{for } 1 \leq k \leq n, \ \sum_k p_k \leq m, \ \text{and} \ \sum_k kp_k = q\}.
$$

Finally, we wish to discuss the algebraic construction corresponding to the adjunction of CW complexes.

Definition **6.7.** Let C be a chain module, let C'' be a chain submodule of the chain module C', and let $f:C''\to C$ be a chain map. The *adjunction chain module* $C\oplus_f C'$ is the quotient of $C\oplus C'$ by the chain submodule generated by the elements $f(c'')-c''$ for $c''\in C''$, together with the induced boundary operator.

THEOREM **6.8.** *Let* $(X,\mathcal{S})$ *and* $(Y,\mathfrak{I})$ *be CW complexes, let* $A\subset X$ *be a subcomplex, and let* $f:A\to Y$ *be a cellular map. Then if* $f_\#:C_*(\mathcal{S}_A)\to C_*(\mathfrak{I})$ *is the map induced by* f, *there is a natural chain equivalence* $C_*(\mathfrak{I}\cup_f \mathcal{S})\to C_*(\mathfrak{I})\oplus_{f_\#}C_*(\mathcal{S})$.

Proof. The reader is referred to section I.7 for the combinatorial structure of the adjunction complex $(Y\cup_f X,\mathfrak{I}\cup_f \mathcal{S})$.

The cellular quotient map $p:Y\cup X\to Y\cup_f X$ of the disjoint union onto the adjunction space induces a natural surjection $\bar{p}:C_*(\mathfrak{I})\oplus C_*(\mathcal{S})\to C_*(\mathfrak{I}\cup_f \mathcal{S})$, where $\bar{p}$ is the composition of the natural isomorphism $e:C_*(\mathfrak{I})\oplus C_*(\mathcal{S})\to C_*(\mathfrak{I}\cup \mathcal{S})$ with $p_\#$. Since $\{\sigma\mid\sigma\in\mathcal{S}\}\cup\{\tau\mid\tau\in\mathfrak{I}\}$ is a basis for $C_*(\mathfrak{I})\oplus C_*(\mathcal{S})$, so is $\{\tau\mid\tau\in\mathfrak{I}\}\cup\{\sigma\mid\sigma\in\mathcal{S}-\mathcal{S}_A\}\cup\{\sigma-f_\#(\sigma)\mid\sigma\in\mathcal{S}_A\}$. But $\{\bar{p}(\tau)\mid\tau\in\mathfrak{I}\}\cup\{\bar{p}(\sigma)\mid\sigma\in\mathcal{S}-\mathcal{S}_A\}$ is a basis of $C_*(\mathfrak{I}\cup_f \mathcal{S})=\operatorname{Im}\bar{p}$, and since $\bar{p}(\sigma-f_\#(\sigma))=0$ for each $\sigma\in\mathcal{S}_A$, it follows that $\operatorname{Ker}\bar{p}=\{\sigma-f_\#(\sigma)\mid\sigma\in\mathcal{S}_A\}$. ∎

Example **6.9.** Let $\mathbf{S}^n$ have a decomposition into cells $\mathcal{S}=\{\sigma^0,\sigma^n\}$, and let D^{n+1} have a decomposition into $\mathfrak{I}=\{\tau^0,\tau^n,\tau^{n+1}\}$ as in I.2.1. If m is an integer, let $f_m:\dot{D}^{n+1}=\mathbf{S}^n\to\mathbf{S}^n$ be a cellular map of degree m, i.e., $[f_m:\tau^n:\sigma^n]=m1\in R$. (We omit a proof of the existence of such a map). Let $X_m=\mathbf{S}^n\cup_{f_m}D^{n+1}$ be the adjunction complex. One easily computes that $C_0(\mathcal{S})\oplus_{f_{m\#}}C_0(\mathfrak{I})\approx R\rho_0$; $C_n(\mathcal{S})\oplus_{f_{m\#}}C_n(\mathfrak{I})\approx R\rho_n$; $C_{n+1}(\mathcal{S})\oplus_{f_{m\#}}C_{n+1}(\mathfrak{I})\approx R\rho_{n+1}$; and that $\partial\rho_n=0$,

$\partial \rho_{n+1} = m\rho_n$. Consequently,

$$H_0(X_m) \approx R;$$

$$H_n(X_m) \approx R/mR;$$

$$H_{n+1}(X_m) \approx {}^mR;$$

$$H_k(X_m) = 0 \quad \text{otherwise.}$$

7. *SEMISIMPLICIAL COMPLEXES*

The procedure by which we pass from the singular complex of a space to the singular chains can be carried through in the case of a general ssc X. We can also pass to the normalized chains. These chain functors are thus given on the category of ssc's. We will show that the cellular chain module of the realization of an ssc X is equivalent to the normalized chain module of X.

Definition **7.1.** Let X be an ssc. We define the *chain module of* X, $C_n(X)$, to be the free R-module generated by X_n. If $f:X \to Y$ is an ss map, define $f:C_n(X) \to C_n(Y)$ by linear extension. We define $d_i:C_n(X) \to C_{n-1}(X)$ and $s_i:C_n(X) \to C_{n+1}(X)$ by linear extension over the basis X_n. Finally, we define $d:C_n(X) \to C_{n-1}(X)$ by linear extension over the basis X_n and $dx = \sum_{i=0}^{n} (-1)^i d_i x$. Let $D(X)$ be the submodule of $C(X)$ generated by degenerate simplexes $s_i x$.

LEMMA **7.2.** (i) *The map d is a differential operator*: $dd = 0$.
(ii) *For any ss map* $f:X \to Y$, $C_*(f) = f:C_*(X) \to C_*(Y)$ *is a chain map.*

Proof. (i) Note that

$$\begin{aligned} ddx_n &= \sum (-1)^{i+j} d_i d_j x_n \\ &= \sum_{i \geq j} (-1)^{i+j} d_i d_j x_n + \sum_{i<j} (-1)^{i+j} d_i d_j x_n \\ &= \sum_{i \geq j} (-1)^{i+j} d_i d_j x_n + \sum_{i<j} (-1)^{i+j} d_{j-1} d_i x_n = 0. \end{aligned}$$

(ii) Since $d_i f(x_n) = f(d_i x_n)$ on each generator of $C_n(X)$, by linear extension we have $df(x_n) = f(dx_n)$. ▮

LEMMA 7.3. (i) *The graded module $D(X)$ is a chain submodule of $C(X)$.*

(ii) *If $f: X \to Y$, then $f(D(X)) \subset D(Y)$. Thus D is a chain functor.*

Proof. (i) It suffices to prove that $d(s_i x) \in D(X)$ for each generator of $D(X)$. But

$$\begin{aligned} d(s_i x_n) &= \sum (-1)^j d_j s_i x_n \\ &= \sum_{j<i} (-1)^i d_j s_i x_n + (-1)^i d_i s_i x_n + (-1)^{i+1} d_{i+1} s_i x_n \\ &\qquad + \sum_{j>i+1} (-1)^i d_j s_i x_n \\ &= \sum_{j<i} (-1)^i s_{i-1} d_j x_n + (-1)^i x_n + (-1)^{i+1} x_n \\ &\qquad + \sum_{j>i+1} (-1)^i s_i d_{j-1} x_n \end{aligned}$$

from the semisimplicial identities. Thus $ds_i x \in D(X)$.

(ii) This statement is obvious. ▮

LEMMA 7.4. *The chain submodule $D(X)$ is acyclic.*

Proof. Define $\Phi_n: D_n(X) \to D_{n+1}(X)$ by the formula

$$\Phi_n(s_k x) = \sum_{j \le k} (-1)^j s_j s_k x$$

and linear extension, where s_k is the degeneracy operator with the largest index for which we can represent $s_k x$. By use of the semisimplicial identities on X we easily establish $d\Phi_n(x) + \Phi_n dx = x$ for $x \in D(X)$. Thus $H_*(D(X)) = 0$. ▮

Definition **7.5.** Let X be an ssc. The chain module $C_*^N(X) = C_*(X)/D(X)$ is the *normalized chain module* of X. If $f:X \to Y$, the map $C_*^N(X) \to C_*^N(Y)$ induced by $C_*(f)$ is the induced map of normalized chain modules.

THEOREM **7.6.** *The map* $q:C_*(X) \to C_*^N(X)$ *induces a natural isomorphism of homology modules.*

Proof. By 7.4, $H_*(D(X)) = 0$. The result follows from considering the exact homology sequence of the short exact sequence of chain modules:

$$0 \to D_*(X) \to C_*(X) \to C_*^N(X) \to 0. \quad \blacksquare$$

If we define $C_*^N(X,A)$ to be $C_*^N(X)/C_*^N(A)$, then q defines a natural transformation $q:C_*(X,A) \to C_*^N(X,A)$ which, by the five-lemma, induces natural isomorphisms in homology.

In dealing with the cellular chains of the realization $|X|$ of an ssc, we observe that: (i) $|X|$ is the quotient of the simplicial complex $M(X)$ (see III.4.6); and (ii) since the cells of $M(X)$ are copies of the standard n-simplex they come with a prescribed orientation, namely the "positive orientation" of the simplex Δ^n with ordered vertices. (This is the homology class in the singular chains of Δ^n mod $\dot{\Delta}^n$ given by the identity map.)

LEMMA **7.7.** *With the orientation described above,* $M(X)$ *is* $\mathcal{R}$*-oriented.*

Proof. The minimal k-cells of $M(X)$ are those which will be identified with k-cells of $|X|$. They are all faces $F^*\Delta^k$ of $\Delta^n \times x_n$ such that Fx_n is a nondegenerate k-simplex x_k of X.

The map which identifies $\Delta^k \times x_k$ with $F^*\Delta^k \subset \Delta^n \times x_n$ is the ordered simplicial map F^*. These minimal cells are identified together when equivalent by order preserving maps and thus if σ, σ', σ'' are any three such, the homeomorphisms $h_{\sigma\sigma'}$, etc., can be chosen so that $h_{\sigma\sigma'}h_{\sigma'\sigma''} = h_{\sigma\sigma''}$. Thus $M(X)$ is $\mathcal{R}$-oriented. $\blacksquare$

THEOREM 7.8. *There is a natural chain isomorphism between the cellular chains of* $|X|$ *and* $C_*^N(X)$ *for any ssc* X.

Proof. Since $\mathfrak{R}$ is an oriented relation, we can take the quotient orientation for $|X|$. Then applying 6.7, each $\epsilon(\sigma,\sigma)=1$. In every equivalence class of minimal n-cells, there is precisely one of the form $\Delta^n \times x_n$, where x_n is nondegenerate, and the elements of this class are mapped onto $|x_n|$. Thus the result follows from 6.7 and 6.8. ∎

We remark that it is possible to construct a geometric realization functor from ssc's to CW complexes in such a way that the cellular chains will be naturally isomorphic to $C_*(X)$, the nonnormalized chains of X. This was done originally by Giever [14] and is described in Hu [19] under "the singular polytope of an ssc." It differs from the one we are studying, due to Milnor, in that the only identifications made are over face operators and not degeneracies. It can also be described as follows. If X is an ssc, let $\bar{X}$ be the following ssc. To each simplex x_n of X, degenerate or not, associate a nondegenerate simplex $\bar{x}_n$ of $\bar{X}$, together with all degeneracies $s_I\bar{x}$. Define $d_i\bar{x}=\overline{d_ix}$ and on degenerate simplexes to satisfy the ss identities. Define $s_i(s_I\bar{x})=s_{I'}\bar{x}$, where $I'=I\cup\{i\}$. Then extend this to a functor by associating with $f:X\to Y$ the map $\bar{f}:\bar{x}\to\overline{f(x)}$. The realization of $\bar{X}$ is the Giever realization of X.

If K is an ssc, there is a natural injection i_K of K into $S(|K|)$: namely, if x is an n-simplex of K and $\varphi_x: \sum(n)\to K$ its characteristic map, then i_K sends x into $|\varphi_x| \in S_n(|K|)$.

PROPOSITION 7.9. *The natural injection* i *induces a natural homology isomorphism.*

Proof. Let $\mathcal{S}$ be the cell structure on the CW complex $|K|$. We identify $C_*(\mathcal{S})$ with $C_*^N(K)$ by the chain isomorphism of 7.8. If $\zeta=\sum a_x x$ is an n-cycle of $C_n(\mathcal{S})=C_n^N(K)$, then from the discussion of 2.17 it follows that $i(\zeta)=\sum a_x|\varphi_x|$ is a representative of $\Psi\{\zeta\}$ in $C_*^N(|K|)$. Thus the map induced in homology by

$\bar{\imath}_K: C_*^N(K) \to C_*^N(| K |)$ determined by i_K is the natural isomorphism $\Psi: H_*(\mathcal{S}) \to H_*(| K |)$, and therefore i_K induces a natural isomorphism of homology. ∎

PROPOSITION **7.10.** *Let X be any space and $j_X: | S(X) | \to X$ the natural map of* III.4.11. *Then $(j_X)_*: H_*(| S(X) |) \to H_*(X)$ is a natural isomorphism.*

Proof. We apply the previous proposition to the case $K = S(X)$. Consider the composition

$$C_*(S(X)) \xrightarrow{(i_{S(X)})_\#} C_*(| S(X) |) \xrightarrow{(j_X)_\#} C_*(X).$$

If x is an n-simplex of the ssc $S(X)$, then $i_{S(X)}$ maps it into $| \varphi_x |$. To find the singular n-simplex of X into which this is mapped by j_X, let t be a point of Δ^n. Then $((j_X)_\#(| \varphi_x |))$ sends t into $j_X(| \varphi_x | (t)) = j_X([t,x]) = x(t)$ (in the notation of III.4.11). That is, $(j_X)_\#(i_{S(X)})_\#$ maps each n-simplex x into the corresponding singular n-simplex of X. Thus this composition induces a natural chain isomorphism of $C_*^N(S(X))$ with $C_*^N(X)$, and therefore induces a natural homology isomorphism. Since $i_{S(X)}$ induces a natural homology isomorphism, it follows that j_X does also. ∎

*8. *REALIZING CELLULAR MAPS*

The results of this section depend heavily on the use of homotopy theory and singular homology, as well as considerable algebraic machinery. Singular homology groups are assumed to have integral coefficients.

Let $(X,\mathcal{S})$ be a CW complex, let A_* be a free chain complex, and let $\varphi: A_* \to C_*(\mathcal{S})$ be a chain map. A *realization* of φ is a pair consisting of a CW complex $(K,\mathfrak{J})$ and a cellular map $f: K \to X$ such that $C_*(\mathfrak{J}) = A_*$ and $f_\# = \varphi: C_*(\mathfrak{J}) \to C_*(\mathcal{S})$.

Our object in this section is to prove the following theorem which is a modification of a theorem of Wall [37].

THEOREM 8.1. *Let $(X,\mathcal{S})$ be a simply connected CW complex, let A_* be a free chain complex with $A_0=Z$, $A_1=0$, and suppose that $\varphi:A_*\to C_*(\mathcal{S})$ is a chain map inducing homology isomorphisms. Then φ has a realization $f:K\to X$.*

COROLLARY 8.2. (*Milnor*). *If X is a simply connected CW complex for which $H_k(X)$ has rank β_k and τ_k torsion coefficients, then X has the homotopy type of a CW complex with $\beta_k+\tau_k+\tau_{k-1}$ cells of dimension k, for $k=1,2,\cdots$.*

COROLLARY 8.3. *If X is a simply connected CW complex for which each $H_k(X)$ has a given free presentation $0\to Z^{r_k}\to Z^{b_k}\to H_k(X)\to 0$ (b_k generators and r_k relations), then X has the homotopy type of a CW complex with b_k+r_{k-1} k-cells for each k.*

Proof. In either case, we can construct a free chain complex A_* with as many generators of A_k as k-cells indicated. By standard techniques one can then construct a chain map $\varphi:A_*\to C_*(X)$ inducing homology isomorphisms. A cellular realization of φ gives the desired homotopy equivalence. ∎

COROLLARY 8.4. *If X is a simply connected space, such that each $H_k(X)$ has a given presentation $0\to Z^{r_k}\to Z^{b_k}\to H_k(X)\to 0$, with $r_1=b_1=0$, then there is a CW complex K, and a map $g:K\to X$ such that K has b_k+r_{k-1} k-cells for each k and g induces homology isomorphisms.*

Proof. We know $j_X:|S(X)|\to X$ induces isomorphisms of homotopy groups in all dimensions, and thus, by the Hurewicz theorem (and the hypothesis of simple connectivity), also of homology groups. By the preceding corollary, we can find $g':K\to |S(X)|$, with K as called for and g' inducing homology isomorphisms. The composition $g=j_Xg'$ satisfies the conditions of 8.4. ∎

The nonsimply connected case is, as usual, more complicated. In [37], Wall proves a result in this direction. Assuming that

$\varphi: A_* \to C_*(X)$ is an isomorphism in dimensions ≤ 2 and induces an isomorphism of homology, there is a realization $f: K \to X$, which, in fact, satisfies $f \mid K^2 =$ identity. It would seem that there are better results possible, but the question is still open.

In the proof of 8.1, it will be necessary to assume the following. If X and A are simply connected, and $H_k(X,A) = 0$ for $1 \leq k \leq n-1$, the Hurewicz homomorphism $\eta: \pi_k(X,A) \to H_k(X,A)$ is an isomorphism for $k \leq n$. One may find a proof in Spanier [35, Chapter 7]. By using this fact and proposition V.1.8, we deduce that for a CW complex X, the group $\pi_n(X^n, X^{n-1})$ is free abelian with basis in one–one correspondence with the n-cells of X.

We will also need the following lemma.

LEMMA **8.5.** *Let X and K be CW complexes with K simply connected, let $f: K^n \to X$ be a cellular map, and suppose that*

$$\xi: \pi_n(K^n, K^{n-1}) \to \pi_n(X^n, X^{n-1})$$

is any homomorphism such that the composition

$$\pi_n(K^n, K^{n-1}) \xrightarrow{\xi} \pi_n(X^n, X^{n-1}) \xrightarrow{i_\#} \pi_n(X, X^{n-1})$$

is the same as $i_\# f_\#$. Then there is a cellular map $g: K^n \to X$ which is homotopic to f rel K^{n-1} and such that $\xi = g_\#$.

Proof. Let $\{\sigma_\lambda^n \mid \lambda \in \Lambda_n\}$ be the set of n-cells of K^n, and let $u_\lambda \in \pi_n(K^n, K^{n-1})$ denote the homotopy class of the characteristic map $\varphi_\lambda: (E^n, \dot{E}^n) \to (K^n, K^{n-1})$ for the cell σ_λ^n. By hypothesis $i_\# \xi(u) = i_\# f_\#(u)$, so that if $h_\lambda: (E^n, \dot{E}^n) \to (X^n, X^{n-1})$ represents $\xi(u_\lambda)$, h_λ and $f\varphi_\lambda$ restrict to homotopic maps on $\dot{E}^n$. We now use the homotopy extension property to extend $f\varphi_\lambda \mid \dot{E}^n$ to a map $\bar{g}_\lambda: E^n \to X^n$ which is homotopic to h_λ. Clearly the map $\bar{g}_\lambda$ defines $g_\lambda: K^{n-1} \cup \sigma_\lambda^n \to X^n$. If g is the union of the maps g_λ, then $g: K^n \to X$ is cellular and is homotopic to f rel K^{n-1}. Since $g_\#(u_\lambda) = \xi(u_\lambda)$ and the elements u_λ are a basis for the free abelian group $\pi_n(K^n, K^{n-1})$, we have $g_\# = \xi$. ∎

We now return to the proof of 8.1. Recall that $(X,\mathcal{S})$ is a simply connected CW complex, A is a free chain complex with $A_0=Z$ and $A_1=0$, and $\varphi:A\to C_*(\mathcal{S})$ is a chain map which induces homology isomorphisms. Let (A^n,∂^n) be the free chain subcomplex of A such that $(A^n)_m=A_m,\partial^n \mid (A^n)_m=\partial \mid A_m$ for $m\leq n$, and $(A^n)_m=0$, $\partial^n \mid (A^n)_m=0$ for $m>n$. Let $\varphi^n:A^n\to C_*(\mathcal{S}^n)$ be the chain map determined by φ. We will realize φ by realizing each φ^n by a complex K^n and a cellular map $f^n:K^n\to X$ such that $f^n \mid K^{n-1}=f^{n-1}$.

The map φ^0 is realized by taking K^0 to be a single vertex σ^0 and taking f^0 to be a map of K^0 into any vertex of X. Since $H_1(X)=0$ and $A_1=0$, we may take $K^1=K^0$ and $f^1=f^0$. Assume inductively that we have realized φ^{n-1} by a cellular map $f^{n-1}:K^{n-1}\to X$. Let K^{n-1} have cells $\mathfrak{J}^{n-1}$, and let

$$(M(f^{n-1}),\mathfrak{M}(f^{n-1}))=(M,\mathfrak{M})$$

be the mapping cylinder complex for $f^{n-1}:K^{n-1}\to X$. In the commutative diagram

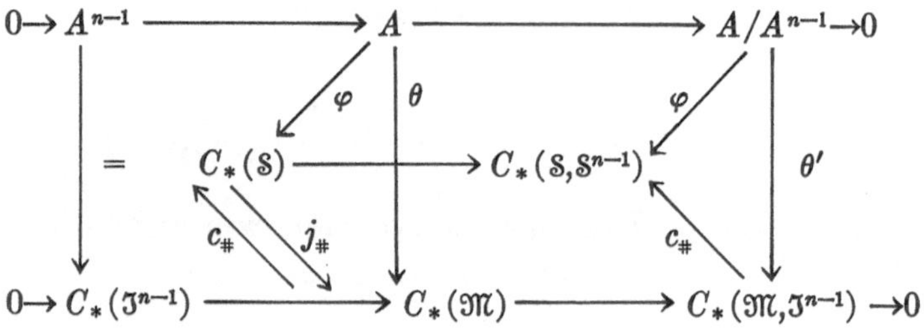

where $j:X\to M$ is the inclusion and $c:M\to X$ is the collapsing map, the rows are short exact sequences of chain complexes, so that there is a map of the homology sequence of (A,A^{n-1}) into that of $(\mathfrak{M},\mathfrak{J}^{n-1})$. Since $\theta_*=j_*\varphi_*$ is an isomorphism, an application of the fives lemma yields the fact that the map θ' induces an isomorphism

$$\theta'_*:H_*(A/A^{n-1})\xrightarrow{\approx}H_*(M,K^{n-1}).$$

Since $f_*^{n-1}:H_q(K^{n-1})\to H_q(X)$ is an isomorphism for $q<n-1$ and an epimorphism for $q\leq n-1$, it follows that $H_q(M,K^{n-1})=0$ for

$q \leq n-1$, and since K^{n-1} is simply connected, the Hurewicz homomorphism $\eta: \pi_n(M,K^{n-1}) \to H_n(M,K^{n-1})$ is an isomorphism. Let θ'' be the composite homomorphism

$$(A^n)_n \to (A^n)_n/\partial A_{n+1} = H_n(A/A^{n-1})$$

$$\xrightarrow{\theta'_*} H_n(M,K^{n-1}) \xrightarrow{\eta^{-1}} \pi_n(M,K^{n-1}).$$

If $\{u_\lambda \mid \lambda \in \Lambda_n\}$ is a basis for $(A^n)_n = A_n$, for each λ, choose a map $h_\lambda: \dot{E}^n \to K^{n-1}$ to represent the homotopy class $\partial_\# \theta''(u_\lambda) \in \pi_{n-1}(K^{n-1})$. Let $\mathcal{E}^n = \bigcup_{\lambda \in \Lambda_n} E^n$, $\dot{\mathcal{E}}^n = \bigcup_{\lambda \in \Lambda_n} \dot{E}^n$, let $h: \dot{\mathcal{E}}^n \to K^{n-1}$ be the map defined by the h_λ, and let $K^n = K^{n-1} \cup_h \mathcal{E}^n$. Then K^n is a CW complex, and by our construction, if $\mathfrak{I}^n$ is the set of cells of K^n, then $C_m(\mathfrak{I}^n) = A_m$ for $m \leq n$.

From the exactness of

$$\pi_n(M,K^{n-1}) \xrightarrow{\partial_\#} \pi_{n-1}(K^{n-1}) \xrightarrow{f^{n-1}} \pi_{n-1}(X)$$

we infer that the composition

$$\dot{E}^n \xrightarrow{h_\lambda} K^{n-1} \xrightarrow{f^{n-1}} X$$

is null homotopic for each λ. Thus for each λ, $f^{n-1} \mid \dot{\sigma}_\lambda{}^n$ extends to a map $\bar{f}_\lambda{}^n: \sigma_\lambda{}^n \to X$, and the union of these maps is a map $\bar{f}^n: K^n \to X$. Our problem is to choose the extension $\bar{f}^n$ properly.

From the remarks preceding lemma 8.5 and definition V.2.1, it follows that $C_n(\mathfrak{I}^n) = H_n(K^n,K^{n-1}) \approx \pi_n(K^n,K^{n-1})$ and $C_n(\mathcal{S}) = H_n(X^n,X^{n-1}) \approx \pi_n(X^n,X^{n-1})$. Thus $A_n \approx \pi_n(K^n,K^{n-1})$. In dimension n, commutativity of the diagram

$$\begin{array}{ccc} A_n & \xrightarrow{\theta'} & C_n(\mathfrak{M},\mathfrak{I}^{n-1}) \\ \downarrow{\scriptstyle \varphi} & & \downarrow{\scriptstyle c_\#} \\ C_n(\mathcal{S}^n,\mathcal{S}^{n-1}) & \xrightarrow{i_\#} & C_n(\mathcal{S},\mathcal{S}^{n-1}) \end{array}$$

implies that the diagram

$$\begin{array}{ccc} A_n = \pi_n(K^n,K^{n-1}) & \xrightarrow{\theta''} & \pi_n(M,K^{n-1}) \\ \Big\downarrow \varphi & & \Big\downarrow c_\# \\ \pi_n(X^n,X^{n-1}) & \xrightarrow{i_\#} & \pi_n(X,X^{n-1}) \end{array}$$

is commutative. Let h_λ' be a characteristic map for the cell σ_λ^n of K^n such that $h_\lambda' \mid \dot{E}^n = h_\lambda$ and let h_λ'' represent $\theta''(u_\lambda)$. Then $ch_\lambda'' \mid \dot{E}^n = f^{n-1}h_\lambda$, and $f^{n-1}h_\lambda$ extends to $\bar{f}_\lambda{}^n$ which is homotopic to ch_λ''. The union of these $\bar{f}_\lambda{}^n$ defines a map $\bar{f}^n: (K^n,K^{n-1}) \to (X,X^{n-1})$, and by the cellular approximation theorem, we can deform $\bar{f}^n$ to a map $f^n: (K^n,K^{n-1}) \to (X^n,X^{n-1})$ such that $f^n \mid K^{n-1} = f^{n-1}$ and $if^n \mid \sigma_\lambda{}^n$ is homotopic to ch_λ''. But then $i_\# f_\#(u_\lambda) = i_\# \varphi(u_\lambda)$, and by lemma 8.5 we have $f_\# = \varphi$. ∎

APPENDIX I

PARACOMPACT SPACES

In this appendix we discuss some of the properties of paracompact spaces, originally defined by Dieudonne [4].

Definition **1.** Let $\mathcal{U} = \{U_\alpha \mid \alpha \in A\}$ be an open covering of a topological space X. A *partition of unity subordinated to* $\mathcal{U}$ is a collection $\{p_\alpha \mid \alpha \in A\}$ of continuous functions $p_\alpha : X \to I$ such that

(i) for each point x of X, there is a neighborhood on which all but a finite number of the p_α vanish;
(ii) $\sum_\alpha p_\alpha = 1$ at each point x of X;
(iii) for each $\alpha \in A$, the function p_α vanishes outside U_α.

A covering $\mathcal{U} = \{U_\alpha \mid \alpha \in A\}$ of X is *numerable* if there is a subordinated partition of unity $p = \{p_\alpha \mid \alpha \in A\}$. The functions p determine a map $p : X \to \Delta(\mathcal{U})$ into the simplex of $\mathcal{U}$, in fact, into the subcomplex $N(\mathcal{U})$, the nerve of $\mathcal{U}$ (see IV.4). As observed in IV.4, the fact that p is a locally finite family of functions implies that each $x \in X$ has a neighborhood V_x such that $\overline{p(V_x)}$ lies in a finite subcomplex of $\Delta(\mathcal{U})$. For V_x we choose a neighborhood of x on which only finitely many functions p_α, say $p_{\alpha_1}, \cdots, p_{\alpha_n}$, are nonzero, so that $p(V_x)$ lies in the simplex with vertices corresponding to $U_{\alpha_1}, \cdots, U_{\alpha_n}$. Thus $p(V_x)$ has a compact closure in $\Delta(\mathcal{U})$, and $p \mid V_x$ is continuous when $\Delta(\mathcal{U})$ is given any topology, such as the CW or metric topology, which reduces to the simplicial topology on its finite subcomplexes.

Definition **2.** A map $p : X \to Y$ is *compact* if the image of p has compact closure in Y and is *locally compact* if each point of X has a neighborhood the image of which has compact closure in Y. A map $p : X \to \Delta(\mathcal{U})$ into a simplex is *full* if the image of p is not contained in any subsimplex.

From the discussion above we have the following.

PROPOSITION 3. *The partitions of unity on X correspond biuniquely to the full locally compact maps of X into CW simplexes $\Delta(\mathfrak{A})$.* ∎

A Hausdorff space X is *paracompact* if each open covering has a subordinated partition of unity.

We note that this is not the usual definition of paracompactness, but is equivalent to it (see proposition 4 below). Also, this definition differs from the one given in Kelley [22] in that it requires X to be Hausdorff.

Examples. 1. Any compact Hausdorff space.
2. Any CW complex (see II.4 and Miyazaki [31]).
3. Any metric space (see Kelley [22, p. 160]).
4. Any Hausdorff Lindelof space (see Dugundji [8, p. 174]). This includes the fact that a Hausdorff σ-compact space is paracompact.

We observe that a paracompact space is normal. Given two closed disjoint sets C and D in X, the complements form an open cover of X. In a subordinated partition of unity $\{p_1, p_2\}$, one function must be 0 on D and 1 on C, and thus must be a Urysohn function for the pair. Since we assume a paracompact space is Hausdorff, it must therefore be normal.

Given a paracompact space X and an open covering $\mathfrak{U} = \{U_\alpha\}$, there are several special types of refinements which we can obtain. These can be constructed easily given a partition of unity subordinated to $\mathfrak{U}$.

PROPOSITION 4. *Let X be a topological space, let $\mathfrak{U} = \{U_\alpha\}$ be an open covering of X, and let $\{p_\alpha\}$ be a partition of unity subordinated to $\mathfrak{U}$. Then*
(i) *$\mathfrak{U}$ has a locally finite open refinement;*
(ii) *$\mathfrak{U}$ has a locally finite closed refinement;*
(iii) *$\mathfrak{U}$ has an open star refinement.*
In each case, these refinements have subordinated partitions of unity.

Proof. The family $\{p_\alpha\}$ of continuous functions determine a map $p: X \to \Delta(\mathfrak{U})$ into the simplex of $\mathfrak{U}$. If $\hat{U}_\alpha$ denotes the vertex corresponding to U_α, then $p_\alpha(x)$ represents the barycentric coordinate corresponding to $\hat{U}_\alpha$. Thus when the simplex is given the metric topology, the map p is continuous, because the barycentric coordinate functions compose with the map p to give continuous functions. The pre-image under p of st $\hat{U}_\alpha$ is an open subset of U_α. The pre-image under p of a locally finite open or closed refinement or of a star refinement is a refinement of the same nature of $\{U_\alpha\}$. Hence we have only to construct a refinement of the type desired for the covering $\{\text{st } \hat{U}_\alpha\}$ of $\Delta(\mathfrak{U})$, and p pulls it back to an appropriate refinement of $\mathfrak{U}$. If we construct such a refinement with a subordinated partition of unity, this will induce one on X with a subordinated partition of unity.

(i) The cover $V_\alpha = \{x \mid b_\alpha(x) > \frac{1}{2} \sup_\beta \{b_\beta(x)\}\}$ is a locally finite open refinement of $\{\text{st } \hat{U}_\alpha\}$, where b_α is the barycentric coordinate function corresponding to $\hat{U}_\alpha$. The subordinated partition of unity is given by:

$$\bar{p}_\alpha(x) = \frac{\max(0, b_\alpha(x) - \frac{1}{2}\sup_\beta \{b_\beta(x)\})}{\sum_\alpha [\max(0, b_\alpha(x) - \frac{1}{2}\sup_\beta \{b_\beta(x)\})]}.$$

The corresponding locally finite open refinement of $\mathfrak{U}$ is $\{U_\alpha^*\}$, where $U_\alpha^* = \{x \mid p_\alpha(x) > \frac{1}{2} \sup_\beta \{p_\beta(x)\}\}$, and the subordinated partition of unity is $\{\bar{p}_\alpha p\}$.

(ii) Take the closures of the sets V_α. Then $\bar{V}_\alpha = \{x \mid b_\alpha(x) \geq \frac{1}{2} \sup_\beta \{b_\beta(x)\}\}$ is a locally finite closed refinement of $\{\text{st } \hat{U}_\alpha\}$.

(iii) Take the barycentric subdivision of $\Delta(\mathfrak{U})$ and apply (i) to the covering by stars. ▮

Lemma 5. (*shrinking lemma*). *In a normal space X, any locally finite open covering $\mathfrak{U}$ has a closed locally finite refinement.*

Proof. Well order the sets U_α of the covering $\mathfrak{U}$. If we have found open sets V_α $(\alpha<\beta)$ such that $\bar{V}_\alpha \subset U_\alpha$ for each $\alpha<\beta$ and $\{V_\alpha \mid \alpha<\beta\} \cup \{U_\alpha \mid \alpha \geq \beta\}$ is an open covering of X, then let $U_\beta^* = U_\beta - \bigcup\{V_\alpha \mid \alpha<\beta\} - \bigcup\{U_\alpha \mid \alpha>\beta\}$. Since U_β is a neighborhood of U_β^* in X, it follows that $U_\beta \supset \bar{U}_\beta^*$. Choose an open set $V_\beta \supset \bar{U}_\beta^*$ whose closure lies in U_β. Then $\{V_\alpha \mid \alpha \leq \beta\} \cup \{U_\alpha \mid \alpha>\beta\}$ is an open covering of X and $\bar{V}_\alpha \subset U_\alpha$ for $\alpha \leq \beta$. The lemma then follows by transfinite induction. Note that since $U_\alpha \supset \bar{V}_\alpha$ and $\{U_\alpha\}$ is locally finite, then $\{V_\alpha\}$ and $\{\bar{V}_\alpha\}$ are also locally finite. ▮

PROPOSITION 6. *A Hausdorff space X is paracompact if and only if each open cover has a locally finite open refinement.*

Proof. The implication "only if" is proved above. To prove the converse, first we prove that a Hausdorff space with the indicated property is normal. This is done by an obvious modification of the proof that a compact (Hausdorff) space is normal, keeping in mind the fact that the union of closures of a locally finite collection of sets is the closure of the union. Next apply the shrinking lemma to any open locally finite refinement of any prescribed open cover $\{U_\alpha\}$ of X. Let $\{V_\alpha\}$ be the resulting closed refinement. For each α let u_α be an Urysohn function 0 off U_α and 1 on V_α. The functions $\{u_\alpha\}$ are a family of nonnegative functions with locally finite supports, hence the sum $Q=\sum u_\alpha$ is everywhere continuous and bounded; since every point lies in some V_α, the sum is positive as well at every point. Therefore $\{u_\alpha/Q\}$ is a family of continuous functions on X. It is clearly a partition of unity subordinated to $\{U_\alpha\}$. ∎

Definition 7. A Hausdorff space X is *perfectly normal* if for each closed subset C, there is a continuous nonnegative real valued function on X for which C is the set of its zeros.

This is a strengthening of the condition of normality; for if C and C' are disjoint closed sets, in a perfectly normal space let f,g vanish on C,C' and nowhere else. Then $u=f(1-g)/\sup(f,g)$ is a Urysohn function for C and C'. Not every normal space is perfectly normal. In Dugundji [8, p. 148] it is proved that a closed set C in a normal space is the set of zeros of a continuous real valued function if and only if it is a G_δ, that is, the intersection of a decreasing family of countably many open sets each containing the closure of the following. Hence in a normal space that does not satisfy the first axiom of countability, some point is a closed set which is not a G_δ. It follows from this criterion that any metric space is perfectly normal. Unlike normality, perfect normality is clearly hereditary; any subspace of a perfectly normal space is again perfectly normal.

The following lemmas lead to a proof of theorem 10.

LEMMA 8. *Every subspace of a space X is paracompact if and only if every open subspace is paracompact.*

Proof. Let X be the space, let A be an arbitrary subspace, and let $\mathfrak{U}=\{U_\alpha\}$ be an open covering of A. Extend each U_α to a set U'_α open in X, and let $A'=\bigcup_\alpha U'_\alpha$. This is an open subset of X, and therefore paracompact by hypothesis. Hence the open cover $\{U'_\alpha\}$ of A' has a subordinated partition of unity which restricts on A to one subordinated to $\mathfrak{U}$. Since X is Hausdorff, the subspace A is also.

The converse is trivial. ∎

Lemma 9. *Let $U=\bigcup_{i\geq 0} K_i$, where each K_i is a closed subset of the interior of K_{i+1}, and suppose each K_i is paracompact. Then U is paracompact (see Michael [24]).*

Proof. Let $\mathcal{W}=\{W_\alpha\}$ be an open covering of U. Let $\mathring{K}_i$ be the interior of K_i. Under our hypothesis, these open sets cover U.

For each i, the closed subset $K_{i+1}-\mathring{K}_{i-1}$ of the paracompact space K_{i+1} is itself paracompact. Set $K_{-1}=\varnothing$, and for each $i\geq 0$ choose a locally finite open refinement $\{V_{i,\beta}\}$ for $\mathcal{W} \mid (K_{i+1}-\mathring{K}_{i-1})$. The collection $\{V_{i,\beta}\cap(\mathring{K}_{i+1}-K_{i-1}) \mid i\geq 0\}$ is a locally finite open refinement of $\mathcal{W}$. Since any two points of U lie in some K_i°, an open Hausdorff subset of U, the space U is Hausdorff and is therefore paracompact. ∎

Theorem 10. *If X is paracompact and perfectly normal, then so is each subspace of X.*

Proof. As we have already observed, perfect normality is hereditary, and therefore we only need to prove that each subset of such a space is paracompact. By lemma 8 we may restrict our attention to the open sets. Since each closed set in a perfectly normal space is a G_δ, each open set is an F_σ, i.e., if U is open in X, $U=\bigcup_i K_i$, where each K_i is closed in the interior of K_{i+1}. However the closed subsets of a paracompact space are paracompact, so that each K_i is paracompact, and by lemma 9, each open set U is paracompact. ∎

APPENDIX II

EXTENSION SPACES AND NEIGHBORHOOD RETRACTS

Throughout this section $\mathcal{C}$ will denote a category of normal topological spaces and continuous maps.

Definition 1. An *absolute retract in* $\mathcal{C}$, an AR($\mathcal{C}$), is a space of $\mathcal{C}$ such that whenever it is embedded in another space of $\mathcal{C}$ as a closed subset, it is a retract of the containing space. An *absolute neighborhood retract in* $\mathcal{C}$, an ANR($\mathcal{C}$), is a space of $\mathcal{C}$ which, whenever embedded in another space of $\mathcal{C}$ as a closed subset, is a retract of a neighborhood in the containing space. If $\mathcal{C}$ is the category of all normal spaces, then Hanner [17] has proposed that these be written ARN and ANRN, respectively. If $\mathcal{C}$ is the category of separable metric spaces, then they are traditionally denoted by AR and ANR, respectively.

The property of being an ANR($\mathcal{C}$) is often derived from the following related property.

Definition 2. An *extension space for* $\mathcal{C}$, an ES($\mathcal{C}$), is a space X, not necessarily in $\mathcal{C}$, such that for any space Y of $\mathcal{C}$, any closed subset A of Y, and any continuous map $f:A\to X$, there is an extension F of f to all of Y. If f extends only over a neighborhood of A in Y, we say that X is a *neighborhood extension space for* $\mathcal{C}$, an NES($\mathcal{C}$).

We will first give some examples and then describe the relationship between these concepts in proposition 7 (also see Hanner [16,17]).

The simplest example of an ES for all normal spaces is the unit interval $I=[0,1]$. This is the content of the following famous result, called Tietze's extension theorem.

THEOREM 3. *For any normal space X, closed subset A of X, and continuous map $f: A \to I$, there is a continuous extension over X.*

A proof may be found in Dugundji [8]. ∎

We recall that a topological vector space is locally convex if there is a neighborhood basis at the origin consisting of convex sets. Any vector space whose topology is derived from a norm (or family of norms) is locally convex, for the vectors of norm less than ϵ form a convex set for each $\epsilon > 0$.

A second theorem of this type is the following.

THEOREM 4. (*Dugundji* [7, 8]). *Any convex subset of a locally convex vector space is an ES for the category of all metric spaces.* ∎

Given some examples of ES($\mathcal{C}$), one can construct others and also some NES($\mathcal{C}$) by using the following proposition.

PROPOSITION 5. (i) *An open subset of an $ES(\mathcal{C})$ is an $NES(\mathcal{C})$.*

(ii) *A retract (neighborhood retract) of an $ES(\mathcal{C})$ is an $ES(\mathcal{C})$ (an $NES(\mathcal{C})$).*

(iii) *The product of any collection (any finite collection) of $ES(\mathcal{C})$ (of $NES(\mathcal{C})$) is an $ES(\mathcal{C})$ (an $NES(\mathcal{C})$).*

(iv) *The union of two relatively open $ES(\mathcal{C})$ ($NES(\mathcal{C})$) whose intersection is also an $ES(\mathcal{C})$ (an $NES(\mathcal{C})$) is an $ES(\mathcal{C})$ (an $NES(\mathcal{C})$). Hence the same result holds for a locally finite union of relatively open sets provided that all the intersections are $ES(\mathcal{C})$ (are $NES(\mathcal{C})$).*

(v) *If B is a locally compact n-ad, all the subspaces are closed; if Y is an $ES(\mathcal{C})$ n-ad (an $NES(\mathcal{C})$ n-ad), which means that the subspaces are all closed and $ES(\mathcal{C})$ ($NES(\mathcal{C})$); and if $\mathcal{C}$ is closed with respect to cartesian multiplication by locally compact spaces, which means that for any n-ad X in $\mathcal{C}^n$, $B \times X$ is an m-ad in $\mathcal{C}$ for suitable m, then the function space Y^B with the compact open topology is an $ES(\mathcal{C})$ (an $NES(\mathcal{C})$).*

Proof. Parts (i)–(iv) are obvious. Statement (v) follows easily from the fact [8, p. 261] that if B is locally compact, a map $X \to Y^B$ is continuous if and only if the associated map $B \times X \to Y$ is continuous. ∎

Definition 6. We say the category $\mathcal{C}$ is *closed under adjunctions* if for any $f: A \to Z$ in $\mathcal{C}$, where $i: A \to X$ is a closed injection map of $\mathcal{C}$, the adjunction space $Z \cup_f X$ is in $\mathcal{C}$. We say $\mathcal{C}$ is an ES category if any space X of $\mathcal{C}$ can be embedded as a closed subspace of an ES($\mathcal{C}$) by a map in $\mathcal{C}$.

The following proposition describes the relation between the concepts of ANR($\mathcal{C}$) and NES($\mathcal{C}$).

PROPOSITION 7. *An ES*($\mathcal{C}$) *in* $\mathcal{C}$ *is an AR*($\mathcal{C}$) *and an NES*($\mathcal{C}$) *in* $\mathcal{C}$ *is an ANR*($\mathcal{C}$). *If* $\mathcal{C}$ *is closed under adjunctions or is an ES category, then an AR*($\mathcal{C}$) *is an ES*($\mathcal{C}$) *and an ANR*($\mathcal{C}$) *is an NES*($\mathcal{C}$).

Proof. If X is an ES($\mathcal{C}$) and is embedded as a closed subset of a space Z of $\mathcal{C}$, then $i: X \to X$, being a map of $\mathcal{C}$ from a closed subspace of a space of $\mathcal{C}$ to X, has an extension over all Z. This extension is a retraction of Z to X. Similarly, if X is a NES($\mathcal{C}$) embedded in Z as a closed subset by a map of $\mathcal{C}$, then there is a neighborhood extension of $i: X \to X$ say to $i': U \to X$. This extension defines a retraction of the neighborhood U of X to X.

If $\mathcal{C}$ is an ES category and X is an ANR($\mathcal{C}$), then X embeds in an ES($\mathcal{C}$) as a closed subset by a map of $\mathcal{C}$. By definition of AR($\mathcal{C}$), this ES($\mathcal{C}$) retracts to X and therefore X is an ES($\mathcal{C}$). If X is an ANR($\mathcal{C}$), then X embeds as a closed subset of an ES($\mathcal{C}$) and some neighborhood U of X retracts to X. Since X is a neighborhood retract in an ES($\mathcal{C}$), it is a NES($\mathcal{C}$).

If $\mathcal{C}$ is closed under adjunctions, and $f: B \to X$ is a map of a closed subset of the space Z, then $Z \cup_f X$ is in $\mathcal{C}$ and contains X as a closed subset. If X is an AR($\mathcal{C}$), there is a retraction r of the adjunction space to X, and this defined an extension of f to all Z. Similarly, if X is an ANR($\mathcal{C}$), there is a neighborhood retraction $r: U \to X$ which defines a neighborhood extension of f over U. ∎

PROPOSITION 8. *The category of all normal spaces and continuous maps is closed under adjunctions. The following are ES categories: all separable metric spaces; all metric spaces; and all compact spaces.*

Proof. Let $f:B\to X$ be a map of a closed subset of the normal space Y into X. Let F_1,F_2 be closed disjoint subsets of the adjunction space. Let U_1,U_2 be open disjoint neighborhoods in X of the sets $F_1\cap X$, $F_2\cap X$. Let V_i be open disjoint subsets of Y containing $f^{-1}(\bar{U}_i)\cup(F_i\cap(Y-B))$. Let $V_i'=(V_i-B)\cup f^{-1}(U_i)$ and V_i'' be the corresponding sets in the image of Y in the adjunction space. Then $U_i\cup V_i''$ are easily seen to be neighborhoods in the adjunction space separating the sets F_i.

If $\mathcal{C}$ is the category of compact spaces, then any cube I^S lies in $\mathcal{C}$ and is an ES($\mathcal{C}$) (in fact is solid). Any compact space embeds in a cube, and its embedding is necessarily a closed map since a cube is Hausdorff.

If X is a metric space, let $B(X)$ be the Banach space of all continuous real valued functions on X with the uniform norm. For each point x let f_x be the function $f_x(y)=d(x,y)$. The mapping $x\to f_x$ embeds X isometrically as a subset of $B(X)$. Let X' be the set of all linear combinations of functions f_x: this contains the convex hull of the image of X, is a convex subset of $B(X)$, and therefore by Dugundji's theorem, an ES for the category of metric spaces. If X is separable, so is X'. We have only to show, therefore, that X is closed in X'. If X^* is the completion of X, observe that $B(X^*)$ is isometric to a convex subset of $B(X)$ which contains X^* and thus also X, and that the map $X^*\to B(X^*)$ is an isometry onto a closed subset of $B(X^*)$. Hence if f_{x_i} is a sequence of points of the image of X in $B(X)$ converging to a point of X' not in X, then the corresponding points x_i of X converge to a point x^* of X^*. Since f_{x^*} is in X', in $(X^*)'$ we have a relation $f_{x^*}=$ a linear combination of points f_x of X. Since x^* is not in X, this is a nontrivial linear relation. But the points of X^* (the functions $d(x^*,\cdot)$) are easily seen to be linearly independent. ∎

Corollary 9. *An ANR($\mathcal{C}$) ($\mathcal{C}=$ metric or separable metric) space X is a retract of an open subset of a normed linear space (namely, X' above).* ∎

Corollary 10. *An ANR is an ANR (metric).*

Proof. An ANR can be embedded in a separable normed linear space, which, by Dugundji's theorem, is an ES (metric). Hence an ANR is an NES (metric). Since it is a metric space it is thus an ANR (metric), by proposition 7. ∎

The following relations among the ANR($\mathcal{C}$) and NES($\mathcal{C}$) conditions are sometimes useful. For proofs, see Hanner [16, 17].

Metric Spaces. A metric space is an ANRN if and only if it is an ANR and an absolute G_δ. That is, it is an NESN if and only if it is a separable NES (separable metric) and an absolute G_δ. This last condition means that there is a topologically equivalent complete metric [22]. This is the case, for example, if the space is compact or locally compact.

A metric space is an ANR (paracompact) if and only if it is an ANR (metric) and an absolute G_δ.

In either of these cases, such a metric space must have the homotopy type of a CW complex (see IV.6.3).

Metric Simplicial Complexes. A simplicial complex with the metric (coarse) topology is an ANR (metric). For a simplex is a convex subset of a locally convex vector space R^S, and is therefore an ES (metric). Any subcomplex is a neighborhood retract, either by a modification of the argument given in II.6.1 for CW complexes or by direct reference to the normal neighborhood theorem [10].

A metric simplicial complex is an absolute G_δ if and only if it contains no infinite dimensional simplex. Hence from the results above for metric spaces:

(i) a metric simplicial complex is an ANR (paracompact) if and only if it contains no infinite dimensional simplex;
(ii) a metric simplicial complex is an ANRN if and only if it has a countable number of vertices and contains no infinite dimensional simplex.

CW Complexes. A CW simplicial complex is an NES (metric).

A CW complex is an ANRN and an ANR if it is locally compact (see [39]).

Bibliography

1. J. F. Adams, An example in homotopy theory, *Proc. Camb. Phil. Soc. Math. Phys. Sci.* **53** (1957), 922–923.
2. M. Barratt, Simplicial and semisimplicial complexes, mimeographed lecture notes, Princeton, 1956.
3. G. DeRham, Complexes à automorphismes et homeomorphie differentiable, *Ann. Inst. Fourier Grenoble* **2** (1950), 51–67.
4. J. Dieudonne, Une generalization des espaces compacts, *J. Math. Pures Appl.* **23** (1944), 65–76.
5. A. Dold, Relations between ordinary and extraordinary homology, *Coll. on Algebraic Topology*, Aarhus 1962, 2–9.
6. C. H. Dowker, Topology of metric complexes, *Amer. J. Math.* **74** (1952), 555–577.
7. J. Dugundji, An extension of Tietze's theorem, *Pac. J. Math.* **1** (1951), 353–367.
8. J. Dugundji, *Topology*, Allyn and Bacon, Inc., Boston, 1966.
9. E. Dyer, Cohomology theories, mimeographed lecture notes, Chicago, 1963.
10. S. Eilenberg and N. Steenrod, *Foundations of algebraic topology*, Princeton University Press, 1952.
11. S. Eilenberg and J. A. Zilber, Semi-simplicial complexes and singular homology, *Ann. of Math.* **51** (1950), 499–513.
12. R. H. Fox, On homotopy type and deformation retracts, *Ann. of Math.* **44** (1943), 40–50.
13. S. Gersten, A product formula for Wall's obstruction, *Amer. J. Math.* **88** (1966), 337–346.
14. J. B. Giever, On the equivalence of two singular homology theories, *Ann. of Math.* **51** (1950), 178–191.
15. B. I. Gray, Spaces of the same n-type for all n, *Topology* **5** (1966), 241–243.
16. O. Hanner, Retraction and extension of mappings of metric and nonmetric spaces, *Ark. Math.* **2** (1952), 315–360.
17. O. Hanner, Some theorems on absolute neighborhood retracts, *Ark. Math.* **1** (1951), 389–408.
18. P. J. Hilton, An introduction to homotopy theory, *Cambridge Tracts in Mathematics, No. 43*, Cambridge University Press, New York, 1953.

19. Sze-Tsen Hu, *Homotopy theory*, Academic Press Inc., New York, 1959.
20. W. Hurewicz and H. Wallman, *Dimension theory*, Princeton University Press, 1948.
21. D. Kan, Functors involving css complexes, *Trans. Amer. Math. Soc.* **87** (1958), 330–346.
22. J. L. Kelley, *General topology*, D. Van Nostrand Co., Inc., Princeton, 1955.
23. W. S. Massey, *Algebraic topology: an introduction*, Harcourt, Brace and World, Inc., New York, 1967.
24. E. Michael, A note on paracompact spaces, *Proc. Amer. Math. Soc.* **4** (1953), 831–838.
25. E. Michael, Continuous selection I, *Ann. of Math.* **63** (1956), 361–382.
26. C. E. Miller, The topology of rotation groups, *Ann. of Math.* **57** (1953), 90–114.
27. J. Milnor, Construction of universal bundles II, *Ann. of Math.* **63** (1956), 430–436.
28. J. Milnor, The geometric realization of a semisimplicial complex, *Ann. of Math.* **65** (1957), 357–362.
29. J. Milnor, On spaces having the homotopy type of a CW complex, *Trans. Amer. Math. Soc.* **90** (1959), 272–280.
30. J. Milnor, On axiomatic homology theory, *Pac. J. Math.* **12** (1962), 337–342.
31. H. Miyazaki, Paracompactness of CW complexes, *Tohoku Math. J.* **4** (1952), 309–313.
32. J. C. Moore, Algebraic homotopy theory, mimeograph lecture notes, Princeton, 1956.
33. D. S. Rim, Modules over finite groups, *Ann. of Math.* **69** (1959), 700–712.
34. J. P. Serre, Modules projectifs et espaces fibris à fibre vectorielle, *Seminaire Dubriel, Algebre et theorie des nombres*, Paris, 1958.
35. E. H. Spanier, *Algebraic topology*, McGraw-Hill Book Co., New York, 1966.
36. N. Steenrod, Cohomology operations, *Annals of Mathematics Studies No. 50*, Princeton University Press, 1962.
37. C. T. C. Wall, Finiteness conditions on CW complexes, *Ann. of Math.* **81** (1965), 56–69.
38. G. W. Whitehead, Generalized homology theories, *Trans. Amer. Math. Soc.* **102** (1962), 227–283.
39. J. H. C. Whitehead, Note on a theorem due to Borsuk, *Bull Amer. Math Soc.* **54** (1948), 1125–1132.

40. J. H. C. Whitehead, On the homotopy type of ANR's, *Bull. Amer. Math. Soc.* **54** (1948), 1133–1145.
41. J. H. C. Whitehead, Combinatorial homotopy I, *Bull. Amer. Math. Soc.* **55** (1949), 213–245.
42. J. H. C. Whitehead, A certain exact sequence, *Ann. of Math.* **52** (1950), 51–110.
43. I. Yokota, On the homology of classical Lie groups, *J. Inst. Poly., Osaka City Univ.* **8** (1957), 93–120.
44. G. S. Young, A condition for the absolute homotopy extension property, *Amer. Math. Monthly* **71** (1964), 896–897.

Index